Fields and Rings

Chicago Lectures in Mathematics

fields and rings

Second Edition

Irving Kaplansky

The University of Chicago Press
Chicago and London

Chicago Lectures in Mathematics Series
Irving Kaplansky, *Editor*

The Theory of Sheaves, by Richard G. Swan (1964)
Topics in Ring Theory, by I. N. Herstein (1969)
Fields and Rings, by Irving Kaplansky (1969; 2nd ed. 1972)
Infinite Abelian Group Theory, by Phillip A. Griffith (1970)
Topics in Operator Theory, by Richard Beals (1971)
Lie Algebras and Locally Compact Groups, by Irving Kaplansky (1971)
Several Complex Variables, by Raghavan Narasimhan (1971)
Torsion-Free Modules, by Eben Matlis (1972)

The University of Chicago Press, Chicago 60637
The University of Chicago Press, Ltd., London

ISBN: 0-226-42450-2 (clothbound); 0-226-42451-0 (paperbound)
Library of Congress Catalog Card Number: 72-78251

CONTENTS

257462

PREFACE TO THE SECOND EDITION

In the second (1970) impression of the first edition, some typographical slips were corrected. Still more are corrected in the present second edition, and I trust the process is converging. I have also added a new section entitled "Notes", and I hope these miscellaneous comments will be useful to readers.

I. Kaplansky

Chicago

PREFACE TO THE FIRST EDITION

These lecture notes combine three items previously available from Chicago's Department of Mathematics: <u>Theory of Fields</u>, <u>Notes on Ring Theory</u>, and <u>Homological Dimension of Rings and Modules.</u> I hope the material will be useful to the mathematical community and more convenient in the new format.

A number of minor changes have been made; these are described in the introductions that precede the three sections.

One point should be noted: the theorems are numbered consecutively within each section. Since there are no cross-references between the sections, no confusion should result.

I trust the reader will not mind a lack of complete consistency, e.g., in Part II the modules are right and the mappings are placed on the right, while in Part III both get switched to the left.

I am very grateful to Mr. Fred Flowers for a fine typing job, and even more for the many excellent suggestions he made during the typing.

I. Kaplansky

Chicago

x

PART I. FIELDS

Introduction

These notes on fields were written in the early 1960's after I had lectured several times on Galois theory. One objective I had in mind was to carry several topics through to a reasonable depth; another was to indicate how one actually goes about computing field degrees and Galois groups.

The foundation of the subject (i. e. the mapping from subfields to subgroups and vice versa) is set up in the context of an absolutely general pair of fields. In addition to the clarification that normally accompanies such a generalization, there are useful applications: to infinite algebraic extensions (§13), and to the Galois theory of differential equations (see my Introduction to Differential Algebra, Hermann, 1957). There is also a logical simplicity to the procedure: everything hinges on a pair of estimates of field degrees and subgroup indices. One might describe it as a further step in the Dedekind-Artin linearization of Galois theory.

1. Field Extensions

Let K be any field and L a field containing K. Then L can be thought of as a vector space over K. The dimension of this vector space is called the dimension (or degree) of L over K and written $[L:K]$. We say that L is finite-dimensional or infinite-dimensional over K according as $[L:K]$ is finite or infinite.

Suppose we have a "tower" of three fields $K \subset L \subset M$. Then three dimensions can be formed and they are connected by a useful relation.

THEOREM 1. Let K, L, M be fields with $K \subset L \subset M$. Then $[M:K]$ is finite if and only if both $[M:L]$ and $[L:K]$ are finite and in that case $[M:K] = [M:L][L:K]$.

Proof. Suppose first that $[M:K]$ is finite. Since L is a subspace of M (as a vector space over K), $[L:K]$ is also finite. Any finite set that spans M over K, for instance a basis, will also span M over L; hence $[M:L]$ is also finite.

To complete the proof of the theorem we assume that $[L:K] = m$, $[M:L] = n$ and we prove that $[M:K]$ is finite and equals mn. Let $u_1, \ldots, u_n$ be a basis of M over L and $v_1, \ldots, v_m$ a basis of L over K. We claim that the mn elements $u_i v_j$ $(i = 1, \ldots, n; j = 1, \ldots, m)$ form a basis of M over K. We must show (1) they span M, (2) they are linearly independent over K.

(1) Let z be any element of M. We can write $z = \Sigma b_i u_i$ with b_i in L. Each b_i can in turn be written $b_i = \Sigma c_{ij} v_j$ with c_{ij} in K. This yields $z = \Sigma c_{ij} u_i v_j$, this last sum being over both i and j $(i = 1, \ldots, n; j = 1, \ldots, m)$.

(2) Suppose $\Sigma c_{ij} u_i v_j = 0$ where each c_{ij} is in K and the sum is over both i and j. We must show $c_{ij} = 0$. Write $b_i = \Sigma c_{ij} v_j$.

Then b_i is in L and $\Sigma b_i u_i = 0$. Since the u's are linearly independent over L, each b_i is 0, i.e., $\Sigma c_{ij} v_j = 0$. Since the v's are linearly independent over K, we conclude that $c_{ij} = 0$.

Let M be any field and S a subset of M. There is clearly a unique smallest subfield of M containing S, namely the intersection of all subfields of M which contain S. We are especially interested in this construction in the case where S consists of a subfield K of M together with one additional element u in M. We then write $K(u)$ for the field in question.

We distinguish two cases.

Case I. There exists no polynomial f with coefficients in K (other than the polynomial identically zero) such that $f(u) = 0$. In this case we say that u is <u>transcendental</u> over K. It is evident that $K(u)$ is the field of all rational functions in u (quotients of polynomials in u) with coefficients in K, where u behaves exactly like an indeterminate over K.

Case II. There does exist a polynomial f with coefficients in K such that $f(u) = 0$. In this case we say that u is <u>algebraic</u> over K. The main facts concerning $K(u)$ are given in the following theorem.

THEOREM 2. <u>Let K be a field,</u> u <u>an element of a larger field, and suppose that</u> u <u>is algebraic over</u> K. <u>Let</u> f <u>be a monic polynomial with coefficients in</u> K <u>of least degree such that</u> $f(u) = 0$, <u>and let this minimal degree be</u> n. <u>Then</u>:

    (a) f <u>is unique</u>,

    (b) f <u>is irreducible over</u> K,

    (c) $1, u, u^2, \ldots, u^{n-1}$ <u>form a vector space basis of</u> $K(u)$ <u>over</u> K,

    (d) $[K(u):K] = n$,

    (e) <u>A polynomial g with coefficients in</u> K <u>satisfies</u> $g(u) = 0$ <u>if and only if</u> g <u>is a multiple of</u> f.

Proof. (a) If f_0 is another monic polynomial of degree n satisfying $f_0(u) = 0$, write $f_1 = f - f_0$. Then $f_1(u) = 0$ and f_1 has degree less than n. If $f_1 \neq 0$, this contradicts our minimal choice of f (of course a harmless multiplication by an element of K will make f_1 monic). Hence $f_1 = 0$, $f = f_0$.

(b) If $f = f_0 f_1$ where f_0, f_1 are polynomials of lower degree with coefficients in K, then either f_0 or f_1 has u as a root and again we have contradicted the minimal choice of f.

(c) The existence of a linear relation with coefficients in K among $1, u, \ldots, u^{n-1}$ implies $g(u) = 0$ with g a polynomial of degree less than n. Hence $1, u, \ldots, u^{n-1}$ are linearly independent over K. We must further show that they span $K(u)$. Write T for the vector subspace of $K(u)$ spanned by $1, u, \ldots, u^{n-1}$. If we show that T is a field it will follow that $T = K(u)$. First we note that T contains every power u^k of u. This is true at least up to $k = n-1$. Suppose it is true for k-1:

(1) $$u^{k-1} = \alpha_0 + \alpha_1 u + \ldots + \alpha_{n-1} u^{n-1} .$$

Multiply (1) by u and recall that u^n is a linear combination of $1, u, \ldots, u^{n-1}$ as a consequence of the equation $f(u) = 0$. We conclude that u^k is in T. It is now clear that T is a ring (indeed an integral domain since it is contained in the field $K(u)$). We must show that any non-zero element z in T has an inverse in T. We can write $z = h(u)$ where h is a non-zero polynomial of degree less than n. Since f is irreducible and h has degree smaller than the degree of f, the greatest common divisor of f and h must be 1. Hence there exist polynomials r and s such that $rf + sh = 1$. Set the variable equal to u in this equation. The result is $h(u) s(u) = 1$ and thus $s(u)$ is the desired inverse of $z = h(u)$.

(d) is an immediate corollary of (c).

(e) If g is not a multiple of f then (since f is irreducible) the greatest common divisor of f and g is 1, and there exist polynomials r and s with rf + sg = 1. Setting the variable equal to u yields a contradiction.

We shall say that a field L containing K is <u>algebraic</u> over K if every element of L is algebraic over K; otherwise we say that L is <u>transcendental</u> over K. If L is a finite-dimensional extension of K it is obvious that L is algebraic over K: if [L:K] = n and u ∈ L then any n + 1 powers of u are linearly dependent over K and this yields a polynomial satisfied by u. It is possible for infinite-dimensional extensions of a field K to be algebraic over K.

We shall often describe K(u) as the field obtained by adjoining u to K; note that so far such adjunction is being discussed only when u is handed to us as an element of a larger field containing K. The dimension of K(u) over K will be called the <u>degree</u> of u over K, and the polynomial f of Theorem 2.2 will be called the irreducible polynomial for u over K. If u is an explicitly given element over an explicit field (usually the rational numbers) the problem of finding the degree of u can in principle be solved by locating the irreducible polynomial for u. Occasionally there are subtler methods that work, and some of these will be developed later in this chapter.

Let L and M be two subfields of a field N. There is a unique smallest subfield of N containing L and M, namely the intersection of all subfields containing L and M. We write L ∪ M for this field. Note that L ∪ M contains the <u>set-theoretic</u> union of L and M but it is usually larger (in fact, L ∪ M is the set-theoretic union only in the trivial case where one of L, M contains the other). If the dimensions of L and M over an underlying field K are known we can get some partial information on the dimension of L ∪ M over K.

THEOREM 3. <u>Let</u> L, M <u>be subfields of a field</u> N <u>and suppose that</u> L <u>and</u> M <u>both contain the field</u> K. <u>Write</u> [L: K] = m, [M: K] = n, [L ∪ M: K] = t.

    (a) t <u>is finite if and only if both</u> m <u>and</u> n <u>are finite,</u>

    (b) <u>In that case</u> t <u>is a multiple of</u> m <u>and of</u> n, <u>and</u> t ≤ mn.

    (c) <u>If</u> m <u>and</u> n <u>are relatively prime,</u> t = mn.

    <u>Proof</u>. If t is finite, so are m and n, since L and M are subfields of L ∪ M. We assume henceforth that m and n are finite. We shall prove t is finite and at most mn by induction on n. The case n = 1 (M = K) being trivial, we assume n > 1. Let u be an element in M but not in K. Write r for the degree of u over K and s for the degree of u over L. We have s ≤ r, for the irreducible polynomial for u over K is a multiple of the irreducible polynomial for u over L. By Theorem 1, [L(u): K] = ms and hence by Theorem 1 again, [L(u): K(u)] = ms/r. Also [M: K(u)] = n/r by Theorem 1. We apply our inductive assumption to the fields L(u) and M over K(u), and deduce [L(u) ∪ M: K(u)] ≤ mns/r² ≤ mn/r. But L(u) ∪ M evidently is the same field as L ∪ M. Hence finally

$$[L \cup M: K] = [L \cup M: K(u)][K(u): K] \leq (mn/r)r = mn.$$

    We have now proved parts (a) and (b) except for the statement that m and n divide t; but this is immediate from Theorem 1. Part (c) is an easy purely number-theoretic consequence of (b).

    The union M ∪ N of two fields assumes a more explicit form when M and N have the form M = K(u), N = K(v). We then write M ∪ N = K(u, v). It is useful to observe that K(u, v) may be thought of in three ways: (1) the smallest subfield (of the given larger field) containing K, u, and v; (2) the result of adjoining v to K(u); (3) the result of adjoining u to K(v).

Exercises

1. If $[L:K]$ is prime, prove that there are no fields properly between K and L.

2. If the degree of u over K is odd, prove that $K(u) = K(u^2)$.

3. Let u be a root of the irreducible polynomial $x^n - a$ over K and suppose that m divides n. Prove that the degree of u^m over K is n/m. What is the irreducible polynomial for u^m over K?

4. Let L be a field algebraic over K and T an integral domain containing K and contained in L. Prove that T is a field.

5. Let L, M be two fields lying between K and N. Let T be the set of all sums $\Sigma y_i z_i$ for $y_i \in L$, $z_i \in M$.

(a) Prove that T is an integral domain,

(b) Prove that $L \cup M$ is the quotient field of T,

(c) If L and M are algebraic over K then $T = L \cup M$ and is algebraic over K.

6. Let u and v be algebraic over K, of degree m and n respectively. Show that u has degree m over $K(v)$ if and only if v has degree n over $K(u)$, and that both statements hold if m and n are relatively prime.

7. Suppose that M and N are finite-dimensional over K and $[M \cup N:K] = [M:K][N:K]$. Prove that $M \cap N = K$. Prove that the converse holds if $[M:K]$ or $[N:K]$ is 2. Give an example where $M \cap N = K$, $[M:K] = [N:K] = 3$, but $[M \cup N:K] < 9$. (Hint: take a real and a non-real cube root of 2).

2. Ruler and compass constructions

We shall indicate briefly in this section how Theorem 1 suffices to show the impossibility of the classical ruler and compass constructions.

To do this, we must translate the geometric problem into algebra. We take the point of view of analytic geometry, labelling the points of the Euclidean plane with ordered pairs of real numbers. We take it as our starting point that all points with integral coordinates are in our possession. We are then allowed to perform ruler and compass constructions to acquire new points. Any point obtainable this way we may call constructible. We call the real number a constructible if the point (a, 0) is constructible. Evidently (a, b) is a constructible point if and only if a and b are constructible numbers.

The ruler and compass constructions that are permitted may be set forth carefully as follows:

(1) Given four distinct points A, B, C, D such that AB and CD are distinct non-parallel lines we are allowed to acquire the point of intersection of AB and CD.

(2) Given distinct points A, B and distinct points C, D such that the circle Γ with center A and radius AB meets CD, we are allowed to acquire the points of intersection of Γ and CD. (The case where two intersecting circles are drawn can be reduced to (1) and (2).)

Now suppose the coordinates of the points A, B, C, D lie in a subfield K of the field of real numbers. Then simple arguments from analytic geometry show that in case (1) the coordinates of the new point lie in K, while in case (2) the coordinates of the new point lie either in K or in $K(\sqrt{a})$ where a is a positive number in K. It follows that any constructible number u lies in a subfield K_n of the real numbers which is the end product of a series of adjunctions

$$\text{Rationals} = K_o \subset K_1 \subset \ldots \subset K_n$$

with each K_i equal to $K_{i-1}(\sqrt{a_i})$, a_i a positive number in K_{i-1}. By iterated use of Theorem 1, $[K_n:K_o]$ is a power of 2; and then by another application of Theorem 1, $[K(u):K_o]$ is a power of 2. We have proved

THEOREM 4. Any constructible real number is algebraic over the rational numbers, and its degree over the rational numbers is a power of two.

It is now a simple matter to demolish the three classical problems on ruler and compass constructions.

(1) Squaring the circle. This means constructing π. Since π is not even algebraic over the rational numbers (this is a hard theorem!) the question of degree does not even enter.

(2) Duplication of the cube. The number $2^{1/3}$ (the real cube root of 2) is to be constructed. Since $x^3 - 2$ is the irreducible polynomial for $2^{1/3}$, $2^{1/3}$ has degree 3 over the rationals and is not constructible.

(3) Trisection of angles. Some angles (e.g. 90°) can be trisected by ruler and compass. We exhibit one angle, 60°, that cannot be trisected by ruler and compass. The question is equivalent to the constructibility of $\cos 20^\circ$ or $u = 2\cos 20^\circ$. From the trigonometric identity $\cos 3\theta = 4\cos^3\theta - 3\cos\theta$ we deduce $u^3 - 3u - 1 = 0$. Since the polynomial $x^3 - 3x - 1$ is irreducible over the rational numbers, it follows that u has degree 3 and is not constructible.

3. Foundations of Galois Theory

It was Galois's remarkable discovery that many questions concerning a field are best studied by transforming them into group-theoretical questions in the group of automorphisms of the field. Usually we are interested in the structure of a field M relative to a subfield K and so it is natural to form the relative group of automorphisms, consisting of those automorphisms of M leaving every element of K fixed. We shall mostly deal with the case where M is finite-dimensional over K, but it is interesting and enlightening to push as far as possible the general case where K and M are absolutely arbitrary.

DEFINITION. Let M be any field, K any subfield. The Galois group of M over K is the group of all automorphisms of M that leave every element of K fixed (in brief: automorphisms of M/K).

In this definition we are taking for granted the evident fact that the automorphisms of M/K do form a group (a subgroup of the full group of automorphisms of M).

Examples. 1. If $M = K$, the Galois group consists just of the identity.

2. K = reals, M = complexes. The Galois group is of order two, consisting of the identity and complex conjugation.

3. K = rationals, $M = K(\sqrt{2})$. Again the Galois group is of order two.

4. K = rationals, $M = K(u)$, u the real cube root of 2. An automorphism of M is determined by what it does to u, and u must be sent into some root of $x^3 - 2$. But the other two roots of $x^3 - 2$ are non-real, and M consists just of real numbers. Hence the Galois group of M/K is the identity. This simple example

shows that the Galois group of M/K can be the identity even when $M \neq K$. See exercise 7 for an example where M is "much bigger than" K and still the Galois group is the identity.

Let M be any field, K any subfield, G the Galois group of M over K. We proceed to set up the fundamental correspondence between subgroups of G and fields lying between K and M. Let L be any intermediate field. We define L' (manifestly a subgroup of G) to be the set of automorphisms of M leaving every element of L fixed. Note that L' is simply the Galois group of M over L. Let H be any subgroup of G. We define H' (manifestly a field between K and M) to be the set of all elements of M left fixed by every automorphism in H. It is natural to call H' the fixed sub-field of M under H. Pictures such as in Figure 1 are helpful in visualizing the two maps.

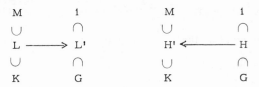

Figure 1

Here (and throughout) we are simply writing 1 for the identity sub-group of G.

Let us try out the priming maps on the four corners of the diagram in Figure 1. In three cases the result is evident and fits the picture: $M' = 1$, $1' = M$, $K' = G$. But it is not necessarily the case that $G' = K$. G' will be the field (say K_o) consisting of all elements of M left fixed by any automorphism fixing K elementwise, and K_o may be properly larger than K. For instance in Example 4 above K_o is actually all of M.

In the favorable case where K_o is equal to K we shall say that M is normal over K. Let us repeat this important definition in different words: M is normal over K if for any u in M but not in K there exists an automorphism of M leaving every element of K fixed but actually moving u. If we are given a field M which is not normal over K, we shall often replace the base field by the larger field K_o (for it is evident that M is normal over K_o; a generalization of this fact will be proved in a moment).

Let us change our notation for the big field from M to N, leaving room for two intermediate fields L and M. We write H and J for two typical subgroups of G, the Galois group of N over K. As an immediate consequence of the definition of the priming operation we have:

(2) $L \subset M$ implies $L' \supset M'$; $J \subset H$ implies $J' \supset H'$.

We next contemplate the result of priming twice. It is obvious that

(3) $L'' \supset L$, $H'' \supset H$.

It may well happen that the field L'' is strictly larger than L. Indeed the assertion $K'' = K$ is just another way of saying that N is normal over K. At any rate we wish to single out the fields for which $L = L''$ and we give them a name: an intermediate field or subgroup will be called closed if it is equal to its double prime. The double prime of any object will be called its closure.

We push one step further still, examining the triple prime. It turns out that nothing new is obtained. Indeed it is a purely formal consequence of (2) and (3) that $L' = L'''$. First, $L''' \supset L'$ by (3) applied to L'. Then start with $L'' \supset L$ and apply (2), obtaining $L''' \subset L'$. Thus $L''' = L'$ for any intermediate field L, and similarly $H''' = H'$ for any subgroup H. In short, any primed object is closed.

If we select any <u>closed</u> intermediate field L and pass to L'
we get a closed subgroup. Priming this closed subgroup returns
us to L. The same thing happens when we take a closed subgroup
and prime twice. We have proved:

THEOREM 5. <u>Let</u> K <u>and</u> N <u>be any fields,</u> K $\subset$ N, <u>and</u> G <u>the</u>
<u>the Galois group of</u> N/K. <u>Then the priming operation sets up a</u>
<u>one-to-one correspondence between the closed subgroups of</u> G
<u>and the closed fields lying between</u> K <u>and</u> N.

Theorem 5 is virtually useless until we collect some information that can tell us which fields or subgroups are closed. In the
next two theorems we prove two estimates on dimensions and these
enable us to prove that closure is at any rate stable under "finite
increases."

THEOREM 6. <u>Let</u> K $\subset$ L $\subset$ M $\subset$ N <u>be fields with</u> $[M:L] = n$
$< \infty$. <u>Then</u> $[L':M'] \leq n$.

<u>Proof.</u> We argue by induction on n, the case $n = 1$ being
trivial. If there exists a field L_o properly between L and M,
then we know $[L':L_o'] \leq [L_o:L]$ and $[L_o':M'] \leq [M:L_o]$. Since
relative field dimensions and group indices are both multiplicative we obtain $[L':M'] \leq [M:L]$. We may therefore assume
that there are no fields between L and M. Necessarily M has
the form L(u). Write f for the irreducible polynomial for u
over L ; f has degree n.

Consider a right coset C of M' in L'. It has the form
$C = M'T$ for some T in L'. Since every automorphism in M'
leaves u fixed, the entire coset C has the same effect on u,
sending u into uT. If $C_o = M'T_o$ is a second right coset distinct from C, then uT_o must be different from uT. For if
$uT_o = uT$, then $T_o T^{-1}$ leaves u fixed, hence leaves $M = L(u)$
elementwise fixed, hence lies in M' ; but then $M'T_o = M'T$.

Note that each uT is a root of f, for T leaves the coefficients of f fixed. Hence the number of right cosets of M' in L' is at most equal to the number of roots of f, which in turn is at most n.

THEOREM 7. Let G be the Galois group of N/K. Let $H \supset J$ be subgroups of G with $[H:J] = n < \infty$. Then $[J':H'] \leq n$.

Proof. Let $C = JT(T \in H)$ be a right coset of J in H. Let x be any element in J'. Then x is left fixed by any automorphism in J. It follows that x is sent into xT by any automorphism in C, and we can unambiguously write $xC = xT$ and speak of applying C to any element of J'. We shall do so in the proof that follows.

We suppose that on the contrary $[J':H'] > n$. Pick $u_1, \ldots, u_{n+1}$ in J' linearly independent over H'. Let $C_1, \ldots, C_n$ denote the right cosets of J in H. We form the equations

$$a_1(u_1 C_1) + a_2(u_2 C_1) + \ldots + a_{n+1}(u_{n+1} C_1) = 0$$
$$a_1(u_1 C_2) + a_2(u_2 C_2) + \ldots + a_{n+1}(u_{n+1} C_2) = 0$$

(4) $\qquad \ldots$

$$a_1(u_1 C_n) + a_2(u_2 C_n) + \ldots + a_{n+1}(u_{n+1} C_n) = 0 .$$

We regard these as n equations for the $n+1$ unknowns $a_1, a_2, \ldots, a_{n+1}$. All the coefficients lie in the field N and so there exists in N a non-trivial solution (i.e. a solution where not all the a's are 0). Among all such solutions pick one with as many zeros as possible; by a harmless change of notation we may assume that this solution has the form

$$a_1, \ldots, a_r, 0, \ldots, 0$$

where each a_i is not zero. We may also assume $a_1 = 1$ (multiply by a_1^{-1}). It is not possible that all the a's lie in H'; for one of the cosets, say C_1, is J itself, and in the first of equations (4) we have $u_i C_1 = u_i$ so that the u's would be linearly dependent over H'. Suppose for definiteness that a_2 is not in H'. Then there is some automorphism S in H such that $a_2 S \neq a_2$. Apply S to the equations (4); the result is $\Sigma (a_i S)(u_i C_j S) = 0$ $(j = 1, \ldots, n+1)$. But $C_1 S, \ldots, C_n S$ are simply a permutation of the cosets $C_1, \ldots, C_n$; hence the new equations are a permutation of the old and $1, a_2 S, \ldots, a_r S, 0, \ldots 0$ is also a solution of (4). Subtracting the two solutions yields a solution with more zeros, non-trivial since $a_2 - a_2 S \neq 0$. This contradiction proves the theorem.

THEOREM 8. (a) <u>Let</u> $K \subset L \subset M \subset N$ <u>be fields. Assume</u> L <u>is closed and that</u> $[M:L] = n < \infty$. <u>Then</u> M <u>is also closed; moreover</u> $[L':M'] = n$.

(b) <u>Let</u> $H \subset J$ <u>be subgroups of the Galois group of</u> N/K. <u>Assume that</u> H <u>is closed and that</u> $[J:H] = n < \infty$. <u>Then</u> J <u>is also closed; moreover</u> $[H':J'] = n$.

<u>Proof.</u> (a) By Theorems 6 and 7,

(5) $$[M'':L''] \leq [L':M'] \leq [M:L] = n.$$

By hypothesis, $L'' = L$. If M'' contains M properly, then the left entry in (5) is larger than n, a contradiction. Hence M is closed. Moreover $[L':M']$ is trapped in the middle of (5) and must also be n.

(b) The proof is essentially the same.

We record an immediate corollary of Theorem 8.

COROLLARY 9. <u>Let</u> G <u>be the Galois group of</u> M <u>over</u> K. <u>Then:</u>

(a) <u>All finite subgroups of</u> G <u>are closed,</u>

(b) If M is normal over K and L is an intermediate field with [L:K] finite then M is normal over L.

We turn to the setup of classical Galois theory, with M finite-dimensional and normal over K and G the Galois group. By Theorem 8 or Corollary 9 all intermediate fields are closed and all subgroups are closed. Hence:

THEOREM 10. (Fundamental theorem of Galois theory) Let M be a normal finite-dimensional extension of K, G the Galois group of M/K. Then there is a one-to-one correspondence between the subgroups of G and the fields between K and M, implemented by the priming operation. In this correspondence the relative dimension of two intermediate fields equals the relative index of the corresponding subgroups. In particular, the order of G is equal to [M:K].

To conclude this section we mention the point of view stressed by Artin: taking the top field as the fundamental object and constructing the bottom field as the fixed subfield under a finite group of automorphisms. The proof is easy (if the results above are used) and we leave it to the reader.

THEOREM 11. Let G be a finite group of automorphisms of a field M and let K be the fixed subfield of M under G. Then M is normal and finite-dimensional over K and the full Galois group of M/K is G.

A nice illustration is to take $M = F(x_1, \ldots, x_n)$ where F is any field and the x's are indeterminates, and G is the group of automorphisms of M obtained by permuting the x's (G is of course isomorphic to the symmetric group S_n on n letters). The fixed field K is the field of all symmetric rational functions in the x's with coefficients in F. We thus exhibit, a little artificially, a pair of fields having S_n as Galois group. Since any finite group is isomorphic to a subgroup of some S_n, we can further exhibit any finite group as a Galois group.

Exercises

1. Let $K \subset L \subset M$ be fields with L normal over K, M normal over L. Assume that any automorphism of L/K can be extended to M. Prove that M is normal over K.

2. In the notation of this section, prove that $(L \cup M)' = L' \cap M'$; $(H \cup J)' = H' \cap J'$. Extend to arbitrary (even infinite) unions. Hence, or otherwise, show that any intersection of closed fields or subgroups is closed.

3. Let K be any infinite field, $M = K(x)$ where x is an indeterminate. Prove that M is normal over K. (Hint: the mapping $x \to x+a$, $a \in K$, induces an automorphism of M/K. If the rational function f/g lies in the fixed field let $h(x, y) = f(x)g(x+y) - g(x)f(x+y)$. Argue that h vanishes for every x and y, hence is identically 0 (this uses the assumption that K is infinite). Deduce that f/g is a constant.)

4. Let K be any field, $M = K(x)$ where x is an indeterminate. Let L be an intermediate field other than K. Prove that M is finite-dimensional over L. (If $r = f/g \in L$, the element x satisfies the equation $rg(x) - f(x) = 0$).

5. Let K be an infinite field, $M = K(x)$ with x an indeterminate, and G the Galois group of M over K. Prove that the only closed subgroups of G are its finite subgroups and G itself.

6. Let K be the field of rational numbers, $M = K(x)$ with x an indeterminate. Prove that the field $K(x^2)$ is closed but the field $K(x^3)$ is not closed.

7. Let K be the field of rational numbers, M the field
of real numbers. Show that the Galois group of M/K is the
identity. (Note that an automorphism preserves order and trap
a given real number between suitable rational numbers).

8. Let $K \subset L \subset M$ be fields with $[L:K] = n$. Show that
there are at most n different isomorphisms of L/K into a subfield of M, an isomorphism of L/K being one that leaves K
elementwise fixed. (Argue as in Theorem 6. If L = K(u), note
that u must go into another root of its irreducible polynomial.
In the general case insert intermediate fields).

9. Let $K \subset L \subset M$ be fields with M normal over K
and $[L:K] = n < \infty$. Prove that any isomorphism of L/K into a subfield of M can be extended to an automorphism of M.
(Note that L' has index n and that its cosets correspond to the
distinct actions on L of automorphisms of M. Use exercise 8).

4. Normality and Stability

In studying a field L lying between K and M we have con-
centrated attention on the property of being closed, that is, of M
being normal over L. We have not asked when it happens that L
is normal over K. Nor have we asked which fields, in the correspondence of Theorem 10, are paired with normal subgroups of G.
It is of course not a coincidence that the same word "normal" is
used in both contexts. But an investigation without any finiteness
assumptions reveals that normality of a subgroup is instead paired
with a stronger property of an intermediate field which we shall call
stability. Of course in the finite-dimensional case it will turn out
that stability of L and normality of L over K coincide.

DEFINITION. Let $K \subset L \subset M$ be fields. We say that L
is stable (relative to K and M) if every automorphism of M/K
sends L into itself.

In this definition we have required only that an automorphism
T of M/K send L into itself, but actually it is automatic that T
sends L onto itself. For there is an inverse automorphism T^{-1}
which (if L is stable) must also send L into itself. Then for
any $x \in L$ we have $xT^{-1} \in L$, $xT^{-1}T = x$ so that T maps onto L.

THEOREM 12. Let G be the Galois group of M/K.

(a) If L is a stable intermediate field, then L' is a normal subgroup of G.

(b) If H is a normal subgroup of G, then H' is a stable
intermediate field.

Proof. (a) Given S in G and T in L' we must show that
STS^{-1} lies in L'. That is, given x in L we must prove
$xSTS^{-1} = x$ or its equivalent $xST = xS$. But this is true since x
lies in L and L is stable, whence xS lies in L.

(b) The proof is essentially the same, read backwards. Given any x in H' and S in G we must prove xS in H'. That is, we must show xST = xS for T in H, or its equivalent $xSTS^{-1} = x$. But this is true since x is in H' and STS^{-1} is in H.

COROLLARY 13. The closure of a normal subgroup is normal; the closure of a stable intermediate field is stable.

Proof. In each case apply Theorem 12 twice.

THEOREM 14. If $K \subset L \subset M$, M is normal over K, and L is stable (relative to K and M), then **L** is normal over K.

Proof. Given an element u in L but not in K we must find an automorphism of L/K that moves u. We know there is an automorphism T of M/K such that $uT \neq u$. Since L is stable T induces an automorphism of L when restricted to L, and this restriction fulfills the requirements.

THEOREM 15. Suppose M is normal over K and f is an irreducible polynomial with coefficients in K having a root u in M. Then f factors over M into distinct linear factors.

Proof. Let $u_1 = u, u_2, \ldots, u_r$ be all the distinct images of u under automorphisms of M/K. Each u_i is a root of f and so we have $r \leq n$ where n is the degree of f. Write $g(x) = (x - u_1) \ldots (x - u_r)$. The coefficients of g are a priori only known to be in M. But any automorphism of M/K merely permutes the u's. Hence the coefficients of g are invariant under every automorphism of M/K and, since M is normal over K, they must lie in K. Now f is the irreducible polynomial for u over K and g is another polynomial over K with u as a root.

By Theorem 2(e), f divides g. Since the degree of g is at most equal to that of f, we deduce g = f. It follows that f, like g, factors over M into a product of distinct linear factors.

THEOREM 16. <u>Let</u> K ⊂ L ⊂ M <u>be fields, and assume that</u> L <u>is normal over</u> K <u>and algebraic over</u> K. <u>Then</u> L <u>is stable.</u>

<u>Proof.</u> Given u in L and an automorphism T of M/K we must prove uT ∈ L. Now u is algebraic over K; let f be its irreducible polynomial over K. By Theorem 15, f factors completely in L. Since uT is a root of f, it must be in L.

THEOREM 17. <u>Let</u> G <u>be the Galois group of</u> M/K, <u>and let</u> L <u>be a stable intermediate field. Then</u> G/L' <u>is isomorphic to the group of all automorphisms of</u> L/K <u>that are extendible to</u> M.

<u>Proof.</u> (Note that by Theorem 12, L' is a normal subgroup of G, so that G/L' is meaningful). Any automorphism T of M/K induces an automorphism of L/K by restricting T to L. This yields a homomorphism from G into the Galois group of L/K. It is clear that the kernel is L' and the image is the set of all automorphisms of L/K that can be extended to M.

Let us return to the classical case where M is finite-dimensional and normal over K. By Theorems 14 and 16 stability of an intermediate field coincides with normality of L over K; furthermore G/L' is the full Galois group of L/K. This follows from the more general result in Exercise 9 of section 3; however it suffices here to note that [G: L'] = [L: K] so that the order of G/L' is the same as the order of the Galois group of L/K. We summarize:

Supplement to Theorem 10.   In the correspondence a field
L is normal over K if and only if the corresponding subgroup H
is normal in G; in this case G/H is the Galois group of L/K .

Exercises

1. Let G be the Galois group of N/K, L and M intermediate fields, H and J subgroups of G.

(a) If M = LT for T ∈ G, then $TM'T^{-1} = L'$.

(b) If $THT^{-1} = J$, then $H' = J'T$.

2. Let G be the Galois group of M/K and L a closed intermediate field. Show that the normalizer of L' in G is the set of all automorphisms of M/K that map L onto itself.

3. Give an example where $K \subset L \subset M$, M is normal over K, L is closed and normal over K, and yet L is not stable. (Take K infinite, M = K(x, y) with x and y indeterminates, and L = K(x).)

5. Splitting Fields

We have established the foundations of Galois theory but we still lack a constructive way of exhibiting fields which are normal over a given field K. To supply this we introduce the concept of a splitting field.

First we need two basic theorems on the existence and uniqueness of the field obtained by adjoining a root of an irreducible polynomial. We shall omit the proofs, but we remark that there are two possible styles for the proof. The first is elementary and explicit: for instance the field $K(u)$ is defined by inventing a symbol u, taking $K(u)$ to be a vector space with basis $1, u, \ldots, u^{n-1}$, and defining multiplication by suppressing multiples of f. Full verification of all the facts is tedious. The second (more sophisticated) method is to define $K(u)$ as the factor ring of the polynomial ring $K[x]$ by the principal ideal (f). This method pushes the tedious details back to the general abstract theory of factor rings.

THEOREM 18. _Let_ f _be an irreducible polynomial with coefficients in a field_ K. _Then there exists a field containing_ K _and a root of_ f.

THEOREM 19. _Let_ K, K_o _be fields and_ S _an isomorphism of_ K _onto_ K_o. _Let_ f _be an irreducible polynomial with coefficients in_ K, f_o _the corresponding polynomial with coefficients in_ K_o. _Let_ $L = K(u)$, $L_o = K_o(u_o)$, _where_ u _and_ u_o _are roots of_ f _and_ f_o _respectively. Then there exists an isomorphism of_ L _onto_ L_o _which coincides with_ S _on_ K _and sends_ u _into_ u_o.

DEFINITION Let f be a polynomial with coefficients in K. We say that M is a __splitting field__ of f over K if f factors completely in M and $M = K(u_1, \ldots, u_r)$ where the u's are all the roots of f.

When there is no need to call attention to the polynomial f,
we shall simply say that M is a splitting field over K. In
Theorem 25 we shall give a criterion for splitting fields that is
independent of the choice of any polynomial.

Note that by Theorem 1 any splitting field over K is finite-
dimensional over K.

THEOREM 20. Let f be any polynomial with coefficients
in K. Then there exists a field M which is a splitting field of
f over K.

Proof. We argue by induction on the degree of f. If f is
linear, $M = K$ will do; more generally if f factors completely
in K, $M = K$. Let, then, g be an irreducible factor of f of
degree greater than one. By Theorem 18 construct $K(u)$ with u
a root of g. Then $f = (x - u)h$, h a polynomial with coefficients
in $K(u)$. It suffices to take M to be a splitting field of h over
$K(u)$ (see Exercise 11).

THEOREM 21. Let K, K_o be fields and S an isomorphism
of K onto K_o. Let f be a polynomial with coefficients in K,
f_o the corresponding polynomial with coefficients in K_o. Let
M be a splitting field of f over K, M_o a splitting field of f_o
over K_o. Then S can be extended to an isomorphism of M
onto M_o.

Proof. We make an induction on $[M:K]$. If $M = K$, then f
factors completely in K, whence f_o factors completely in K_o,
and $M_o = K_o$. We may assume that f has an irreducible factor
g of degree greater than one; let g_o be the corresponding irreducible factor of f_o over K_o. Let u (resp. u_o) be a root
of g (resp. g_o) in M (resp. M_o). By Theorem 19 the isomorphism S can be extended to an isomorphism of $K(u)$ onto

$K_o(u_o)$; we continue to write S for the extended map. Now M is a splitting field of f over $K(u)$ and M_o is a splitting field of f_o over $K_o(u_o)$ -- see Exercise 10. Since $[M:K(u)] < [M:K]$ our inductive assumption shows that S can be extended to an isomorphism of M onto M_o.

At this point in the subject we must cope with a special difficulty that occurs only for characteristic $p \neq 0$: the possibility that an irreducible polynomial may (in a larger field) have a repeated root. As a technical help we introduce, purely formally, the derivative of a polynomial. If $f = \Sigma a_i x^i$ is a polynomial with coefficients in K, we define $f' = \Sigma i a_i x^{i-1}$. By routine computation we can verify that the usual rules for derivatives hold: $(f+g)' = f' + g'$, $(fg)' = f'g + fg'$, $(cf)' = cf'$ for c in K.

THEOREM 22. Let f be a polynomial with coefficients in K, a an element in K. Then the following statements are equivalent: $(x-a)^2$ divides f, $x-a$ divides both f and f'.

Proof. If $f = (x-a)^2 g$, then $f' = (x-a)^2 g' + 2(x-a)g$ is divisible by $x-a$. Suppose $f = (x-a)h$ and $f' = h + (x-a)h'$ is divisible by $x-a$. Then $x-a$ divides h, whence $(x-a)^2$ divides f.

THEOREM 23. Let f be an irreducible polynomial with coefficients in K. The following three statements are equivalent:

(1) In every splitting field of f over K, f factors into distinct linear factors,

(2) In some splitting field of f over K, f factors into distinct linear factors,

(3) $f' \neq 0$.

Proof. (1) implies (2) is obvious.

(2) implies (3). Suppose on the contrary that $f' = 0$. Then for any root a of f, $x - a$ divides both f and f'. Hence (Theorem 22), $(x - a)^2$ divides f, a contradiction.

(3) implies (1). If on the contrary f has a repeated factor $(x - a)^2$ in the splitting field, then $x - a$ divides both f and f'. But f is irreducible over K and f' is a genuine polynomial of lower degree. Hence f and f' have the greatest common divisor 1 and $rf + sf' = 1$ for suitable polynomials r, s with coefficients in K. On setting $x = a$ we get a contradiction.

When can it happen that f' is the zero polynomial ? If $f = \Sigma \, a_i x^i$ is not merely a constant, then $f' = \Sigma \, i a_i x^{i-1}$ will have in it genuine terms unless each term is annulled by the insertion of the coefficient i. So the characteristic must be $p \neq 0$ and each i must be divisible by p. In other words: f must be a polynomial in x^p.

DEFINITION. Let f be an irreducible polynomial over K. We say that f is separable over K if any (hence all) of the statements in Theorem 23 hold. An element u algebraic over K is said to be separable over K if its irreducible polynomial is separable over K. A field L algebraic over K is separable over K if every element is separable over K. To avoid ambiguity we shall not define separability over K of a polynomial unless it is irreducible over K.

We emphasize again that separability is automatic in the case of characteristic 0.

THEOREM 24. Let M be a finite-dimensional extension of K. The following three statements are equivalent:

(1) M is normal over K,

(2) M is separable over K and M is a splitting field over K,

(3) M is a splitting field over K of a polynomial whose irreducible factors are separable.

Proof. (1) implies (2). Let u be an element of M and f its irreducible polynomial over K. By Theorem 15, f factors over M into distinct linear factors. Hence u is separable over K. Since this is true for every u in K, M is separable over K.

Let $v_1, \ldots, v_r$ be a basis of M over K, let f_i be the irreducible polynomial over K for v_i and write $g = f_1 \ldots f_r$. By Theorem 15 again, each f_i factors completely in K and hence so does g. Clearly M is a splitting field of g over K.

(2) implies (3). Say M is a splitting field of f over K, and $f = f_1 \ldots f_r$ is the factorization of f into irreducible factors over K. Each f_i is the irreducible polynomial for an element in M which by hypothesis is separable over K. Hence each f_i is separable over K.

(3) implies (1). Assume that M is a splitting field of f over K where the irreducible factors of f are separable. Let G be the Galois group of M/K. We shall prove that M is normal over K by proving that the order of G is equal to [M:K]. If f factors completely in K, then M = K and there is nothing to prove. Let g be an irreducible factor of f having degree greater than one; say the degree of g is r. Let u be a root of g and write L = K(u), H = L'. Just as in the proof of Theorem 7, we have [G:H] = the number of images of u in automorphisms of M/K. But every one of the r distinct roots of g is such an image, for

if v is another root there is by Theorem 19 an isomorphism S of K(u) onto K(v) leaving K elementwise fixed, and then by Theorem 21, S can be extended to an automorphism of M. Hence [G:H] = r = [L:K]. By induction, the order of H is equal to [M:L], for M is still the splitting field of f over L, and the irreducible factors of f over L are separable (they divide the irreducible factors of f over K). Multiplying, we get that the order of G is [M:K] and hence M is normal over K.

We shall now derive a criterion for splitting fields that does not name any special polynomial.

THEOREM 25. Let L be a finite-dimensional extension of K. The following statements are equivalent:

(1) L is a splitting field over K,

(2) Whenever an irreducible polynomial over K has a root in L it factors completely in L.

Proof. (1) implies (2). Assume that L is a splitting field of f over K, and let g be an irreducible polynomial over K with root u in L. We must show that g factors completely in L. Suppose on the contrary that over L, g has an irreducible factor h of degree greater than one. Adjoin to L a root v of h. Then by Theorem 19 there is an isomorphism S of K(u) onto K(v) which is the identity on K. Now L is a splitting field of f over K(u) and L(v) is a splitting field of f over K(v). By Theorem 21, S can be extended to an isomorphism of L onto L(v). But this is nonsense, for [L(v):K] is strictly larger than [L:K].

(2) implies (1). Let $v_1, \ldots, v_r$ be a basis for L over K, let f_i be the irreducible polynomial for v_i over K, and write $f = f_1 \ldots f_r$. Then f factors completely in L by hypothesis and L is a splitting field of f over K (compare Exercise 11).

THEOREM 26. Let $K \subset L$ be fields, $[L:K]$ finite. There exists a field M containing L such that M is a splitting field over K and no field other than M between L and M is a splitting field over K. If M_o is a second such field, then there is an isomorphism of M onto M_o which is the identity on L. If L is separable, then M is normal over K.

Proof. The construction of M has been foreshadowed in several earlier arguments. We take a basis $v_1, \ldots, v_r$ of L over K, $f = f_1 \cdots f_r$ where f_i is the irreducible polynomial for v_i over K, and then take M to be a splitting field of f over L. Then M is also a splitting field of f over K (compare exercise 11) and it is normal over K if L is separable over K (for then each f_i is seaprable over K). Any splitting field over K which contains L must split each f_i for they each acquire a root in L. This shows that M has the property asserted in the theorem. Any second such field M_o must also be a splitting field of f over K or L, and the uniqueness asserted follows from Theorem 21.

We shall call a field having the properties of M in Theorem 26 a split closure of L over K; if L is separable over K we call M a normal closure of L over K.

In concluding this section we summarize the connection between splitting fields and normality: for characteristic 0, normal is the same as splitting field; for characteristic p, normal is splitting field plus separability.

Exercises

1. Let $K \subset L \subset M$ be fields with L normal over K (possibly infinite-dimensional) and M a splitting field over L of a polynomial with coefficients in K whose irreducible factors over L are separable. Prove that M is normal over K. (Use Theorem 21 and Exercise 1 of §3).

2. If u_i is separable over K $(i = 1, \ldots, r)$ prove that $K(u_1, \ldots, u_r)$ is separable over K.

3. Let f be a polynomial of degree n with coefficients in K. Let L be a splitting field of f over K. Prove that $[L:K]$ is a divisor of $n!$

4. Let K be a field of characteristic $\neq 2, 3$. Show that the following statements are equivalent:

(a) Any sum of squares in K is a square.

(b) Whenever a cubic polynomial f factors completely in K, so does f'. (This problem is motivated by the observation -- an easy consequence of Rolle's theorem -- that (b) holds for any polynomial over the reals. I have been unable to determine just what fields have this property).

5. State and prove the form that Exercise 4 takes for characteristic 2.

6. Let $K \subset L \subset M$ be fields with L a splitting field over K. Prove that L is stable.

7. Suppose that M is a splitting field over K and L is an intermediate field. Prove that L is a splitting field over K if and only if L is stable. Show further that G/L' is the full Galois group of L/K.

8. Let M be a split closure of L over K. Prove that $M = L_1 \cup \ldots \cup L_r$ where L_i is isomorphic to L over K.

9. If $K \subset L \subset M$ and M is separable over K, then M is separable over L.

10. If $K \subset L \subset M$ and M is a splitting field of f over K, then M is a splitting field of f over L.

11. Suppose $K \subset L \subset M$ and L is generated over K by some of the roots of a polynomial f with coefficients in K. Prove that M is a splitting field of f over K if and only if M is a splitting field of f over L.

6. Radical Extensions

In a large part of classical algebra the main theme was the search for "explicit" solutions of equations. While the meaning of "explicit" was perhaps not made precise, it was always clear that rational operations and extractions of roots were permitted. The formula for solving a quadratic equation was already known to the ancients. During the seventeenth century similar (but increasingly complicated) formulas were found for the cubic and quartic equations. The search for an explicit solution of quintic equations ended in defeat when Abel proved that such a formula was impossible. Shortly thereafter Galois proved the same thing in a dramatic new way that truly explained the failure and moreover made it possible to settle, at least in principle, whether a specific equation with numerical coefficients could be solved in the prescribed way. In this section we shall present Galois's results.

First we must put in precise form the field-theoretic meaning of solution by radicals.

DEFINITION. A field L is a <u>radical extension</u> of K if L has the form $K(u_1, \ldots, u_m)$ where some power of u_i lies in $K(u_1, \ldots, u_{i-1})$ for $i = 1, \ldots, m$.

Note that a radical extension of K is clearly finite-dimensional over K.

By inserting further u's , if necessary, we can arrange that in each case a <u>prime</u> power of u_i lies in $K(u_1, \ldots, u_{i-1})$. In the proof of Theorem 27 we shall suppose that his has been done.

We proceed at once to the main theorem on radical extensions. We state and prove it here only for characteristic 0, but it is true for any characteristic (see Exercise 1).

THEOREM 27. If K has characteristic 0, $K \subset L \subset M$ and M is a radical extension of K, then the Galois group of L/K is solvable.

Four lemmas will precede the proof of Theorem 27. In these lemmas we drop the restriction of characteristic 0.

LEMMA 1. The union of a finite number of radical extensions is a radical extension.

Proof. It is enough to do the case of the union of two radical extensions. Suppose then that L, M are radical extensions with $L = K(u_1, \ldots, u_m)$ and $M = K(v_1, \ldots, v_n)$ exhibiting the fact that they are radical extensions. Then

$$L \cup M = K(u_1, \ldots, u_m, v_1, \ldots, v_n)$$

shows that $L \cup M$ is a radical extension of K.

LEMMA 2. If L is a radical extension of K and M is a split closure of L over K, then M is a radical extension of K.

Proof. This is immediate form Lemma 1 and Exercise 8 of §5.

LEMMA 3. Let p be a prime and L a splitting field of $x^p - 1$ over K. Then the Galois group of L/K is abelian.

Proof. If the characteristic is p, then $x^p - 1 = (x-1)^p$, and $L = K$. For characteristic $\neq p$, $x^p - 1$ has distinct roots. Let ϵ be a root different from 1. Then ϵ has multiplicative order p and thus $1, \epsilon, \epsilon^2, \ldots, \epsilon^{p-1}$ are all the roots of $x^p - 1$. Hence $L = K(\epsilon)$. An automorphism of L is determined by what it does to ϵ. Say the automorphisms S and T send ϵ into ϵ^i and ϵ^j respectively. Then ST and TS both send ϵ into ϵ^{ij}. Thus $ST = TS$, and the Galois group of L/K is abelian.

LEMMA 4. Let K be a field in which $x^n - 1$ factors into linear factors. Let a be any element in K, and L a splitting field of $x^n - a$ over K. Then the Galois group of L/K is abelian.

Proof. If u is one root of $x^n - a$ then the general root has the form ϵu where $\epsilon^n = 1$ and so ϵ lies in K. It follows that $L = K(u)$, and that an automorphism of L/K is determined by what it does to u. Say the automorphisms S and T send u into ϵu and ηu respectively (ϵ, η roots of $x^n - 1$ in K). Then ST and TS both send u into $\epsilon \eta u$. Hence the Galois group of L/K is abelian.

Proof of Theorem 27. If K_o denotes the closure of K relative to the Galois group of L/K (i.e., K_o is the fixed subfield of L under the automorphisms of L/K), nothing in the problem is changed if we replace K by K_o (see Exercise 2). Hence we may assume that L is normal over K. If N denotes a normal closure of M over K then, by Lemma 2, N is a radical extension of K. Thus (changing notation again) we may assume that M is normal over K. Since the Galois group of L/K is a homomorphic image of that of M/K, and since a homomorphic image of a solvable group is solvable, we have only to show that the Galois group of M/K is solvable. Thus we may henceforth forget about L.

Let $M = K(u_1, \ldots, u_n)$ be the generation of M that shows M to be a radical extension. We shall argue by induction on n. As noted above, we may assume that u_1^p lies in K for some prime p. Let M_o be a splitting field of $x^p - 1$ over M. Let M_1 be the subfield of M_o generated by K and the roots of $x^p - 1$. The four fields involved are shown in Figure 2.

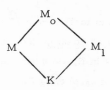

Figure 2

If we show that the Galois group of M_o/K is solvable, it will follow that the Galois group of M/K is solvable, again because a homomorphic image of a solvable group is solvable. Now M_1 is a normal extension of K with a Galois group which is abelian by Lemma 3. Hence it will suffice to show that the Galois group of M_o/M_1 is solvable, for a group is solvable if a normal subgroup and factor group are solvable. Now $M_o = M_1(u_1, \ldots, u_n)$, for M_o is generated from K by the u's and the roots of $x^p - 1$, and the latter are already in M_1. Let G denote the Galois group of M_o/M_1 and H the subgroup corresponding to $M_1(u_1)$ in the Galois correspondence (Figure 3).

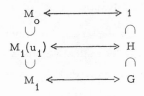

Figure 3

Since $x^p - 1$ factors completely in M_1, $M_1(u_1)$ is a splitting field of $x^p - u_1^p$ over M_1 and hence it is normal with a Galois group which is abelian by Lemma 4. Thus G/H is abelian. To prove that G is solvable it remains finally to show that H is solvable. This follows from our inductive assumption, for M_o is a radical extension of M_1 generated by a chain $u_2, \ldots, u_n$ of n-1 elements. This completes the proof of Theorem 27.

We add a supplemnt to Theorem 27 which refers more directly to the solution of an <u>equation</u> by radicals. If f is a polynomial with coefficients in K, we define the Galois group of f to be the Galois group of a splitting field of f over K -- it follows readily from Theorem 21 that a choice of a different splitting field of f yields the same Galois group up to isomorphism. We usually think of the Galois group of f as a group of permutations on the roots of f, i. e. as a subgroup of the symmetric group S_n on n letters if f has n roots.

THEOREM 28. <u>Let</u> f <u>be an irreducible polynomial over a field</u> K <u>of characteristic</u> 0. <u>Suppose there exists a radical extension</u> L <u>of</u> K <u>which contains a root of</u> f. <u>Then the Galois group of</u> f <u>over</u> K <u>is solvable.</u>

<u>Proof.</u> Enlarge L to a normal closure N of L over K; by Lemma 2, N is still a radical extension of K. N contains a splitting field M of f over K and by Theorem 27 the Galois group of M over K is solvable.

Now we see Galois's explanation of why polynomials of the n-th degree can be solved by radicals up to n = 4 but not in general for $n \geq 5$: the reason is that S_n is solvable for $n \leq 4$ but not for $n \geq 5$.

To get an explicit example of an equation not solvable by radicals we could resort to the device at the end of §3 to construct an extension with Galois group S_n. However the base field for this extension is too peculiar; we would like to have a down to earth example with base field the rational numbers. Perhaps the simplest class of examples is furnished by the following theorem.

THEOREM 29. <u>Let</u> p <u>be a prime and</u> f <u>an irreducible</u>
<u>polynomial of degree</u> p <u>over the rational numbers. Assume that</u>
f <u>has exactly two non-real roots. Then the Galois group</u> G <u>of</u>
f <u>is the full symmetric group</u> S_p <u>on the p roots of</u> f.

Proof. G has order divisible by p, for in obtaining the
splitting field of f we first adjoin an element of degree p. Hence
G has an element of order p, necessarily a p-cycle. Complex
conjugation induces an automorphism which is a transposition on
the roots, for it merely interchanges the two non-real roots.
That $G = S_p$ now follows from an easy lemma on permutation
groups.

LEMMA. <u>Let</u> p <u>be a prime. If a subgroup</u> G <u>of</u> S_p <u>con-</u>
<u>tains a transposition and a p-cycle then</u> G <u>is all of</u> S_p.

Proof. By taking a suitable power of the p-cycle if necessary,
we can arrange the notation so that it is $f = (1\,2\,\ldots\,p)$ and the
transposition is $g = (1\,2)$. We form repeated conjugates: $gfg = f_1$
$= (2\,1\,\ldots\,p)$, $f_1^{-1}gf = g_1 = (1\,3)$, $g_1^{-1}f_1g_1 = f_2 = (2\,3\,1\,\ldots\,p)$,
$f_2^{-1}g_1f_2 = g_2 = (1\,4), \ldots$. This produces all of $(1\,2), (1\,3), \ldots, (1\,p)$
and they generate S_p.

An explicit illustration of Theorem 27 is provided by
$x^5 - 6x + 3$. It is irreducible over the rationals by Eisenstein's
criterion, and a crude inspection of its graph reveals that is has
exactly 3 real roots. Hence the Galois group of $x^5 - 6x + 3$ is
S_5, and it is impossible to express any root of $x^5 - 6x + 3$ by
a formula involving only rational operations and extractions of
n-th roots.

Exercises

1. Prove that Theorems 27 and 28 are valid for any characteristic. (Use Exercise 7 of §5).

2. If $K \subset L \subset M$ and M is a radical extension of K, then M is a radical extension of L.

3. Show that in Lemmas 3 and 4 the Galois group is actually cyclic.

7. The Trace and Norm Theorems

Let L be a normal finite-dimensional extension of K, with Galois group given by $S_1, \ldots, S_n$. For any a in L we define the trace and norm of a :

$$T(a) = aS_1 + aS_2 + \ldots + aS_n ,$$
$$N(a) = (aS_1)(aS_2)\ldots(aS_n) .$$

Clearly $T(a)$ and $N(a)$ lie in K, for they are fixed under all automorphisms of L/K. Trace is additive and norm multiplicative: $T(a+b) = T(a) + T(b)$, $N(ab) = N(a)N(b)$. For $a \in K$ we have $T(a) = na$, $N(a) = a^n$.

It is possible to define trace and norm for any finite-dimensional extension (not necessarily normal); there are certain subtleties and we shall not do it here.

It is our objective in this section to prove two theorems which characterize, in the case of a cyclic Galois group, the elements of trace 0 and norm 1; we shall then give an application for each theorem. The main tool is a fundamental result on linear independence of automorphisms.

THEOREM 30. <u>Any distinct automorphisms of a field</u> K <u>are linearly independent over</u> K.

<u>Remark 1.</u> It is not being assumed that the automorphisms in question form a group or even that one of them is the identity. However since linear dependence is defined by finite sums, it might as well be a finite set of automorphisms.

<u>Remark 2.</u> Linear independence of $S_1, \ldots, S_n$ over K means: if $a_1(xS_1) + a_2(xS_2) + \ldots + a_n(xS_n) = 0$ for every x in K, then all the a's must be 0.

Proof. Suppose on the contrary that $S_1, \ldots, S_n$ are linearly dependent over K. Among all dependence relations pick a "shortest" one, i.e., one with as many zeroes as possible. Say this shortest relation is

$$(6) \qquad a_1(xS_1) + a_2(xS_2) + \ldots + a_r(xS_r) = 0 \qquad \text{(all } x \in K).$$

Of course r must be greater than one. Since S_1 and S_2 are distinct, there exists b in K with $bS_1 \neq bS_2$. In (1) we may replace x by bx, obtaining

$$(7) \qquad a_1(bS_1)(xS_1) + a_2(bS_2)(xS_2) + \ldots + a_r(bS_r)(xS_r) = 0 \qquad \text{(all } x \in K$$

Multiply (6) by bS_1 and subtract (7):

$$a_2(bS_1 - bS_2)(xS_2) + \ldots + a_r(bS_1 - bS_r)(xS_r) = 0 .$$

This is a shorter dependence relation, non-trivial since the coefficient of xS_2 is not zero.

THEOREM 31. Let L be normal over K with a Galois group which is cyclic of order n generated, say, by S. Then an element a in L has trace 0 if and only if it is of the form $b - bS$ for some b in L.

Proof. If $a = b - bS$, then

$$T(a) = a(I + S + S^2 + \ldots + S^{n-1})$$

$$= (b - bS) + (bS - bS^2) + \ldots + b(S^{n-1} - S^n) = 0$$

since $S^n = I$.

Conversely, assume $T(a) = 0$. By Exercise 1 there exists in L an element c with $T(c) = 1$. Define $d_0 = ac$, $d_1 = (a + aS)cS$, and in general

$$d_i = (a + aS + \ldots + aS^i)(cS^i)$$

for $0 \leq i \leq n-2$. Set $b = d_0 + d_1 + \ldots + d_{n-2}$. Since

$d_i S = (aS + aS^2 + \ldots + aS^{i+1})(cS^{i+1})$, we find $d_{i+1} - d_i S$
$= a(cS^{i+1})$ for $0 \leq i \leq n-3$. Also $d_{n-2} S = -a(cS^{n-1})$ since
$T(a) = 0$. Hence

$$b - bS = d_0 + (d_1 - d_0 S) + (d_2 - d_1 S) + \ldots + (d_{n-2} - d_{n-3} S) - d_{n-2} S$$

$$= ac + a(cS) + a(cS^2) + \ldots + a(cS^{n-1})$$

$$= a$$

since $T(c) = 1$.

As an application of Theorem 31 we shall describe the structure of a normal extension of degree p in the case where the characteristic is the same prime p.

THEOREM 32. Let L be normal over K, where $[L:K]$ is a prime p which is also the characteristic of K. Then $L = K(u)$ where u is a root of an irreducible polynomial over K of the form $x^p - x - a$.

Proof. The Galois group of L/K is cyclic of order p, say generated by S. The element 1 satisfies $T(1) = 0$. By Theorem 31, we can write $1 = uS - u$ for some u in L (take u to be the negative of the b in Theorem 31). We have $uS = 1 + u$, hence $u^p S = (1 + u)^p = 1 + u^p$. It follows that $a = u^p - u$ is invariant under S and hence lies in K. Since there are no fields properly between K and L, and u is not in K, we have $L = K(u)$. It follows that $x^p - x - a$ must be the irreducible polynomial for u.

The next two theorems are the "multiplicative" companions of what might be called the additive theory of Theorems 31 and 32. Theorem 33 (in a slightly more special context) was Theorem 90 in Hilbert's 1897 report on algebraic number theory.

THEOREM 33 (Hilbert's "Theorem 90"). <u>Let</u> L <u>be normal</u> <u>over</u> K <u>with a cyclic Galois group generated, say by</u> S. <u>Then an</u> <u>element</u> a <u>in</u> L <u>has norm</u> 1 <u>if and only if it has the form</u> $a = b/bS$ <u>for some</u> $b \neq 0$ <u>in</u> L.

Proof. If $a = b/bS$, then

$$N(a) = a(aS) \ldots (aS^{n-1}) = \frac{b}{bS} \cdot \frac{bS}{bS^2} \cdots \frac{bS^{n-1}}{bS^n} = 1$$

since $S^n = I$.

Suppose conversely that $N(a) = 1$. Write $d_o = ac$, $d_1 = (a \cdot aS)(cS)$ and in general

$$d_i = (a \cdot aS \ldots aS^i)(cS^i)$$

for $0 \leq i \leq n-1$. Note that $d_{n-1} = cS^{n-1}$ since $N(a) = 1$. Note also that $d_{i+1} = a(d_i S)$ for $0 \leq i \leq n-2$. By Theorem 30 there must exist a choice of c in L such that $b = d_o + d_1 + \ldots + d_{n-1}$ is not 0. Then

$$bS = d_o S + d_1 S + \ldots + d_{n-1} S = \frac{1}{a}(d_1 + d_2 + \ldots + d_{n-1}) + cS^n .$$

$S^n = I$, so $cS^n = d_o/a$. Hence $bS = b/a$, as desired.

THEOREM 34. <u>Let</u> L <u>be normal over</u> K <u>with a Galois</u> <u>group which is cyclic of order</u> n, <u>say generated by</u> S. <u>Assume</u> <u>that the characteristic is prime to</u> n <u>and that</u> $x^n - 1$ <u>factors</u> <u>completely in</u> K. <u>Then</u> $L = K(u)$ <u>where</u> u <u>is a root of an irre-</u> <u>ducible polynomial over</u> K <u>of the form</u> $x^n - a$.

Proof. There are n distinct roots of $x^n - 1$ in K and they form a multiplicative group. Any finite multiplicative group in a field is cyclic. Let a generator be ϵ. We have $N(\epsilon) = \epsilon^n = 1$. By Theorem 33 we can write $\epsilon = uS/u$ for suitable u in L. Then $uS = \epsilon u$, $u^n S = \epsilon^n u^n = u^n$. Hence $a = u^n$ is invariant under S and lies in K. If n were prime, we could now conclude just as in Theorem 32 that $L = K(u)$. To cope with the possibility that n is composite we need an additional argument. In any event $K(u)$ is a splitting field of $x^n - a$ over K. The automorphisms $I, S, \ldots, S^{n-1}$ send u into the distinct elements $u, \epsilon u, \ldots, \epsilon^{n-1} u$. Hence $K(u)$ admits n automorphisms over K and $[K(u):K] \geq n$, whence $K(u) = L$, since $[L:K] = n$. It follows that $x^n - a$ must be the irreducible polynomial for u over K.

Theorem 34 is the basic ingredient for establishing a converse to Theorem 27. This time the restriction to characteristic 0 cannot be dropped (but a companion theorem could be proved in which extensions such as those in Theorem 32 are allowed).

THEOREM 35. <u>Let</u> K <u>be of characteristic</u> 0, L <u>a finite-dimensional normal extension of</u> K <u>with a solvable Galois group</u> G. <u>Then</u> L <u>can be embedded in a radical extension of</u> K.

Proof. We make an induction on $[L:K]$. In order to apply Theorem 34 we need the usual technical maneuver to cope with the possibly missing roots of unity. We may assume that G has a normal subgroup H of prime index p. Let N be a splitting field over L of $x^p - 1$. Then N is normal over K and still has a solvable Galois group. Let M be the subfield of N obtained by adjoining to K the roots of $x^p - 1$. Then N is also normal over M. If we prove that N can be embedded in a radical extension of M we are finished, for M is a radical extension of K.

Claim: the Galois group of N/M is isomorphic to a subgroup of G. To see this we map any automorphism T of N/M into its restriction to L (which of course is an automorphism of L/K). Let T be in the kernel of this homomorphism; then T leaves both L and M elementwise fixed and hence is the identity. So the mapping is an isomorphism, as claimed.

Case I. The Galois group of N/M is isomorphic to a proper subgroup of G. Then by our inductive assumption N can be embedded in a radical extension of M.

Case II. The Galois group of N/M is isomorphic to all of G. Let us simply call it G again. Let P be the intermediate field which corresponds to H. Then $[P:M] = p$, P is normal over M, and M contains the p-th roots of 1. By Theorem 34, $P = M(u)$ with u a root of a polynomial $x^p - a$, i.e., P is a radical extension of M. N is normal over P with the solvable Galois group H. By induction N can be embedded in a radical extension of P. The theorem is proved.

Exercises

1. Let L be finite-dimensional and normal over K. Then any element in K is the trace of a suitable element in L.

2. Let K be a field of characteristic p and a, b non-zero element of K.

(a) Prove that $x^p - x - a$ is either irreducible or factors completely in K. (Hint: if u is a root then all the roots are of the form $u + i$, $i = 0, \ldots, p-1$).

(b) Prove that $x^p - b^{p-1}x - a$ is either irreducible or factors completely in K. (Set $x = by$).

(c) Prove that $x^p + b^{p-1}a^{-1}x^{p-1} - a^{-1}$ is either irreducible or factors completely in K. (Set $x = 1/y$).

3. Show that in Theorem 31 the element b is unique up to addition of an element in K.

4. Show that in Theorem 33 the element b is unique up to multiplication by a non-zero element in K.

8. Finite Fields

A finite field is one having only a finite number of elements. We are already familiar with the examples furnished by the fields J_p (integers mod p) for every prime p. In this section we shall determine what further finite fields exist and substantially exhibit their structure.

A finite field K must have characteristic $p \neq 0$ for otherwise it contains a copy of the rational numbers. Suppose $[K:J_p] = n$. Then K has exactly p^n elements. More generally: an n-dimensional vector space over a field with q elements has exactly q^n elements, for expressing the elements of V in terms of a basis we have just q choices for each of the n coordinates and therefore q^n choices in all.

We shall prove that for every power p^n of a prime there does exist a field with p^n elements and that any two such are isomorphic. The key fact is given in the following theorem.

THEOREM 36. _A field_ K _has_ p^n _elements if and only if it is a splitting field over_ J_p _of_ $x^{p^n} - x$.

Proof. Suppose K has p^n elements. The multiplicative group of non-zero elements in K has order $p^n - 1$. Hence $u^{p^n - 1} = 1$ for any $u \neq 0$ in K. Putting this equation in the form $u^{p^n} = u$ we have it satisfied for $u = 0$ as well. Thus $x^{p^n} - x$ has its full quota of p^n distinct roots in K. Since these roots constitute all of K, K is a splitting field of $x^{p^n} - x$ over J_p.

Conversely, suppose K is a splitting field of $x^{p^n} - x$ over J_p. The derivative of $x^{p^n} - x$ is -1 and hence the p^n roots are all distinct. Moreover, since the mapping $u \to u^{p^n}$ preserves addition as well as multiplication, the p^n roots form a field. But since K is generated over J_p by the roots of $x^{p^n} - x$, K must be equal to this field. Hence K has p^n elements.

We know (Theorems 20 and 21) that splitting fields exist and are unique up to isomorphism. Hence:

THEOREM 37. _For any power_ p^n _of a prime_ p _there exists a field with_ p^n _elements and any two such are isomorphic._

We next investigate how the Galois theory of finite fields works out. The result is gratifyingly simple and decisive.

THEOREM 38. _Let_ $K \subset L$ _be finite fields. Then_ L _is normal over_ K _and the Galois group of_ L/K _is cyclic._

Proof. It suffices to treat the case $K = J_p$. For we have $L \supset K \supset J_p$; if it is proved that L is normal over J_p with a cyclic Galois group it will follow that L is also normal over K and that the Galois group is cyclic (being a subgroup of a cyclic group).

That L is normal over J_p is immediate from Theorem 36; L is a splitting field over J_p of a polynomial with distinct roots.

For any field L of characteristic p the mapping S sending every element into its p-th power is at least an isomorphism of L _into_ itself. But when L is finite, S is necessarily onto. If $[L:J_p] = n$, S^n is the identity. No lower power is the identity: if $S^k = I$ with $k < n$ then the polynomial $x^{p^k} - x$ has p^n roots in L, which is impossible in a field. Hence $I, S, \ldots, S^{n-1}$ are all distinct and constitute the whole Galois group of L/J_p .

Exercises

1. Suppose K with p^m elements is contained in L with p^n elements. Prove that n is a multiple of m. Prove further that the Galois group of L/K is generated by the automorphism $x \to x^{p^m}$.

2. Let K be any finite field, n an integer. Prove that there exists an irreducible polynomial over K of degree n.

3. Let f be a polynomial with the property that its roots (in some splitting field) form a field (this is the situation encountered in the proof of Theorem 36). Prove that the characteristic is p and that f has the form $x^{p^n} - x$.

4. Let p, q be distinct primes. Assume that q is a primitive root of p, i.e., for no $d < p-1$ is $q^d - 1$ divisible by p. Prove that $(x^p - 1)/(x-1)$ is irreducible over the field of integers mod q. (Hint: $x^p - 1$ factors completely in the field of q^d elements if and only if p divides $q^d - 1$).

5. For $n \geq 3$, prove that $x^{2^n} + x + 1$ is reducible over J_2. (Hint: if u is a root, raise the equation $u^{2^n} = u + 1$ to the 2^n-th power).

6. Prove: in any finite field any element can be written as the sum of two squares.

7. Let $K \subset L$ be finite fields. Prove that any element of K is the norm of some element of L. (Hint: consider the homomorphism from L^* -- the multiplicative group of non-zero elements of L -- to K^* given by $u \to N(u)$. Find the size of the kernel by using Theorem 33 and Exercise 4 of §7).

9. Simple Extensions

The following theorem gives a neat necessary and sufficient condition for a finite-dimensional extension to be generated by a single element. Such an extension we call simple.

THEOREM 39. Let M be any finite-dimensional extension of K. Then M is a simple extension of K if and only if there are only a finite number of intermediate fields.

Proof. If. Assume first that K is finite. Then M is also finite. The multiplicative group of non-zero elements of M is cyclic. Any generator of this cyclic group will generate M.

Suppose then that K is infinite. Pick an element u in M such that $[K(u):K]$ is as large as possible. We claim $K(u) = M$. Suppose the contrary and pick v in M but not in $K(u)$. As a ranges over K we get a formally infinite list of intermediate fields $K(u + av)$. Two of these must coincide, say $K(u+av)$ and $K(u+bv)$. Then $K(u+av)$ contains $u+av$ and $u+bv$, hence $(a-b)v$, hence v and also u. Thus $[K(u+av):K] > [K(u):K]$, a contradiction.

Only if. We now assume $M = K(u)$ and have to prove that there are only a finite number of intermediate fields. Let f be the (unique monic) irreducible polynomial for u over K. Let L be a typical intermediate field and let g be the monic irreducible polynomial for u over L. Say $g(x) = x^r + a_1 x^{r-1} + \ldots + a_r$. We claim that $L = K(a_1, \ldots, a_r)$. Certainly L contains $K(a_1, \ldots, a_r)$, so that $[M: K(a_1, \ldots, a_r)] \geq r$. On the other hand, u satisfies an equation of degree r over $K(a_1, \ldots, a_r)$ so that the opposite inequality $[M: K(a_1, \ldots, a_r)] \leq r$ holds. This proves that $L = K(a_1, \ldots, a_r)$ and shows that L is uniquely determined by g.

Now there are only a finite number of monic divisors of f
(think of the complete factorization of f in a splitting field and
recall that this factorization is unique). Hence there are only a
finite number of L's.

The criterion of Theorem 39 can be applied almost immediately to the case of a separable extension.

THEOREM 40. Any finite-dimensional separable extension
L of a field K is a simple extension of K.

Proof. Embed L in M, a normal closure of L over K.
By the Galois correspondence it is immediate that there are only
a finite number of fields between K and M. Hence the same is
true between K and L. Apply Theorem 39.

Exercises

1. Prove: if K is infinite and u, v are separable algebraic
over K then $K(u, v) = K(u + av)$ for some a in K. Is this true if
K is finite?

2. Let K have characteristic p and $L = K(u, v)$ where
$u^p, v^p \epsilon K$ and $[L:K] = p^2$. Show that L is not a simple extension
of K and exhibit an infinite number of intermediate fields.

10. Cubic and Quartic Equations

We turn to the following question: how can the Galois group of an equation be found explicitly? We shall give fairly complete results for cubic and quartic equations and the initial step in investigating general equations.

Let f be a polynomial with coefficients in a field K of characteristic $\neq 2$. Let M be a splitting field of f over K, and G the Galois group of M/K. Assume that the roots $x_1, \ldots, x_n$ of f are distinct. We think of G as a group of permutations on $x_1, \ldots, x_n$ and thus as a subgroup of S_n.

Write

$$\Delta = (x_1 - x_2)(x_1 - x_3) \cdots (x_{n-1} - x_n) = \prod_{i < j} (x_i - x_j)$$

and $D = \Delta^2$. Since D is invariant under all the permutations of the x's, we have $D \in K$. We call D the discriminant of f. We know that a permutation of the x's is even if it leaves Δ fixed and odd if it sends Δ into $-\Delta$. If H denotes the subgroup of G consisting of even permutations (the "even subgroup" of G), it follows that H and $K(\Delta)$ correspond in the Galois correspondence between subgroups and intermediate fields. We summarize:

THEOREM 41. Let K be a field of characteristic $\neq 2$. Let f be a polynomial over K, M a splitting field of f. Assume that f has distinct roots in M. Let G be the Galois group of M/K, thought of as a group of permutations of the roots of f. Then in the Galois correspondence $K(\Delta)$ corresponds to the even subgroup of G. In particular, G consists of even permutations if and only if Δ lies in K.

We continue the analysis, now adding the assumption that f is irreducible. It is then true that the order of G is divisible by n. A stronger statement can be made: G is <u>transitive</u> on the x's. That is, for any x_i and x_j there exists an automorphism of M/K sending x_i into x_j (this is immediate from Theorems 12 and 21).

If f is a cubic equation, G is a subgroup of S_3. The only subgroups of S_3 with order divisible by 3 are A_3 and S_3 (they are of course also the only transitive subgroups). Hence:

THEOREM 42. <u>The Galois group of a separable irreducible cubic over</u> K <u>is either</u> A_3 <u>or</u> S_3. <u>For</u> K <u>of characteristic</u> $\neq 2$, <u>it is</u> A_3 <u>if and only if the discriminant is a square in</u> K.

Elementary computation shows that the discriminant of $x^3 + px + q$ is $-4p^3 - 27q^2$. Except for characteristic 3, any cubic can be reduced to this form by a change of variable $x = y + c$.

If the given base field K is a subfield of the real numbers, we can use to good advantage information on the reality of the roots. In fact if only one root is real (and the cubic is irreducible), the Galois group is S_3 as in Theorem 29. In this case the discriminant is negative. If all three roots are real the discriminant is positive, and we must determine whether it is a square in K.

Let f be a quartic with distinct roots x_1, x_2, x_3, x_4. The Galois group G is a subgroup of S_4. Since the crucial normal subgroup of S_4 is V, the subgroup consisting of $(1), (12)(34)$, $(13)(24)$, and $(14)(23)$, it is a very natural step to form the expressions $\alpha = x_1 x_2 + x_3 x_4$, $\beta = x_1 x_3 + x_2 x_4$, $\gamma = x_1 x_4 + x_2 x_3$. Obviously any permutation in V leaves α, β, and γ fixed. Conversely, an easy argument shows that a permutation leaving α, β, γ fixed is necessarily in V. Hence the field $K(\alpha, \beta, \gamma)$ corresponds to $G \cap V$, and $G/(G \cap V)$ is the Galois group of $K(\alpha, \beta, \gamma)$ over K.

The polynomial $(y - \alpha)(y - \beta)(y - \gamma)$ is called the resolvent cubic of f. If $f = x^4 + bx^3 + cx^2 + dx + e$, an elementary computation shows that the resolvent cubic is

$$y^3 - cy^2 + (bd - 4e)y - b^2e + 4ce - d^2 .$$

Now let us assume that the quartic f is irreducible. Then G is a transitive subgroup of S_4. The eligible groups are found to be: S_4, A_4, one of the groups of order 8 (there are three, all conjugate), V, and the cyclic group of order four generated by a 4-cycle. Checking $G/(G \cap V)$ in each case we find there is only one possibility of ambiguity. The details are as follows:

THEOREM 43. Let f be a separable irreducible quartic over K. Let m be the degree over K of the splitting field of the resolvent cubic of f. Let G be the Galois group of f over K. Then

(1) If m = 6, G is S_4,

(2) If m = 3, G is A_4,

(3) If m = 1, G is V,

(4) If m = 2, G is either of order 8 or cyclic of order 4. One way to distinguish the two is to determine whether f is still irreducible after the roots of the resolvent cubic are adjoined.

It remains to explain the final sentence in Theorem 43. The Galois group of f over $K(\alpha, \beta, \gamma)$ is $G \cap V$. If G is of order 8, then $G \cap V = V$ and is still transitive on the roots; hence f is still irreducible over $K(\alpha, \beta, \gamma)$. But if G is cyclic of order 4, then $G \cap V$ has order 2, and f must factor over $K(\alpha, \beta, \gamma)$.

Exercises

1. Let $x^3 + ax + b$ be irreducible over a field K of characteristic 2. Prove that the Galois group is A_3 or S_3 according as $y^2 + by + a^3 + b^2$ has or has not a root in K.

2. Let $x^3 + px + q$ be irreducible over a finite field K. Prove that $-4p^3 - 27q^2$ is a square in K.

3. Let $x^4 + dx + e$ be irreducible over a finite field K of characteristic 2. Prove that d is a cube in K.

4. Let $x^4 + ax^2 + b$ be irreducible over a field K of characteristic $\neq 2$. Let G be the Galois group. Prove:

(1) If b is a square in K, $G = V$,

(2) If b is not a square in K but $b(a^2 - 4b)$ is, G is cyclic of order 4,

(3) If neither b nor $b(a^2 - 4b)$ is a square in K, G has order 8.

5. Let f be a separable irreducible quartic over K, G the Galois group of f, u a root of f. Show that there is no field properly between K and $K(u)$ if and only if $G = A_4$ or S_4.

6. Let K be a subfield of the real numbers, f an irreducible quartic over K. Let G be the Galois group of f. Prove: if f has exactly two real roots, then G is the whole symmetric group S_4 or is of order 8.

7. Let $x^4 + bx^3 + cx^2 + bx + 1$ be irreducible over a field K of characteristic $\neq 2$. Let G be the Galois group of f. Prove:

(1) If $c^2 + 4c + 4 - 4b^2$ is a square in K, $G = V$,

(2) If $c^2 + 4c + 4 - 4b^2$ is not a square in K but $(c^2 + 4c + 4 - 4b^2)(b^2 - 4c + 8)$ is, G is cyclic of order 4,

(3) If neither $c^2 + 4c + 4 - 4b^2$ nor

$(c^2 + 4c + 4 - 4b^2)(b^2 - 4c + 8)$ is a square in K, G is of order 8.

8. Over a field K of characteristic $\neq 2$, let f be a cubic whose discriminant is a square in K. Prove that f is either irreducible or factors completely in K.

9. Over any base field K prove that $x^3 - 3x + 1$ is either irreducible or factors completely in K.

11. Separability

We shall examine in greater detail the behavior of a finite-dimensional extension field with respect to separability. We first define a concept which is, so to speak, the extreme opposite of separability.

DEFINITION. Let K be a field of characteristic p. An element u is _purely inseparable_ over K if for some k, u^{p^k} lies in K. A field L containing K is purely inseparable over K if every element of L is purely inseparable over K.

THEOREM 44. _If an element_ u _is both separable and purely inseparable over_ K _then it lies in_ K.

Proof. Let f be the irreducible polynomial for u over K. Then f has distinct roots (in a splitting field). On the other hand, f is a divisor of a polynomial of the form $x^{p^k} - a$ which has all its roots equal. Hence f is linear and u lies in K.

THEOREM 45. _If_ u _is algebraic over_ K _then_ u^{p^n} _is separable over_ K _for some_ n.

Proof. We argue by induction on the degree of u over K. If u is separable, all is well. Otherwise the irreducible polynomial for u over K is actually a polynomial in x^p, whence u^p has lower degree over K than u does. By induction some p^k-th power of u^p is separable over K, i.e., $u^{p^{k+1}}$ is separable over K.

Now we assemble in a single theorem the major results on the structure of a finite-dimensional extension with respect to separability.

THEOREM 46. <u>Let</u> N <u>be a finite-dimensional extension</u> <u>of</u> K. Then:

(1) <u>There exists a unique largest subfield</u> L <u>separable</u> <u>over</u> K,

(2) <u>There exists a unique largest subfield</u> M <u>purely in-</u> <u>separable over</u> K,

(3) $L \cap M = K$,

(4) N <u>is purely inseparable over</u> L,

(5) N <u>is separable over</u> M <u>if and only if</u> $L \cup M = N$,

(6) <u>If</u> N <u>is a splitting field over</u> K <u>then</u> $L \cup M = N$; <u>also,</u> N <u>is normal over</u> M, L <u>is normal over</u> K, <u>and the Galois groups</u> <u>of</u> N/M <u>and</u> L/K <u>are isomorphic.</u>

Proof. (1) Simply take L to be the set of all elements of N separable. It is immediate from Exercise 2 of §5 that L is a subfield and of course it is the unique largest separable subfield.

(2) Take M to be the set of all purely inseparable elements. It is obvious that M is a subfield with the desired property.

(3) The elements of $L \cap M$ are both separable and purely inseparable over K. By Theorem 44, they must lie in K.

(4) Let u be any element of N. By Theorem 45 some u^{p^n} is separable over K, hence lies in L. This shows that N is purely inseparable over L.

(5) Suppose that N is separable over M. Then evidently N is separable also over $L \cup M$, and by (4) N is purely inseparable over $L \cup M$. Hence $N = L \cup M$.

Conversely, suppose $N = L \cup M$. If $u_1, \ldots, u_r$ are any generators of L over K, then $N = M(u_1, \ldots, u_r)$ and N is separable over M by Exercise 1 of §5. (Actually by Theorem 40 a single u would do, but there is no need to insist on this economy).

(6) Let M_o denote the fixed subfield of N under automorphisms of N/K. We claim that $M_o = M$. First suppose
$u \in M$. Then u satisfies a polynomial equation over K of the
form $x^{p^n} - a$ which has all its roots equal. Hence u cannot be
moved by an automorphism and $u \in M_o$. Suppose $u \in M_o$ and
let f be the irreducible polynomial for u over K. If v is
another root of f then $v \in N$ by Theorem 25, and there is an
automorphism of N/K sending u into v by Theorems 19 and
21. It follows that all the roots of f are equal and then
(see Exercise 3) that u is purely inseparable over K.

We have thus proved $M = M_o$, i.e., N is normal over M.
By (5), $L \cup M = N$. Let T be any automorphism of N/M.
T must send L onto itself for separable elements go into separable elements. By restricting T to L we get an automorphism
of L/K. The resulting homomorphism from the Galois group of
N/M to the Galois group L/K is onto (Theorem 21) and one-to-
one for if T is in the kernel it leaves both L and M elementwise fixed and hence also $N = L \cup M$. Finally, L is normal over
K, since $L \cap M = K$ and only elements of M are fixed under all
automorphisms of N/K. This completes the proof of Theorem 46.

The transitivity of separability is now easily proved.

THEOREM 47. If $K \subset L \subset M$, L is separable over K, and
M is separable over L, (all extensions finite-dimensional), then
M is separable over K.

Proof. Let P denote the maximal separable subfield of M,
regarded as an extension field of K. Of course, $P \supset L$. By (4) of
Theorem 46, M is purely inseparable over P. But M is also
separable over P, since it is separable over L. Hence $M = P$.

We ask: what fields have the property that all extensions are separable? The answer is given by Theorem 48. First we make a definition.

DEFINITION. A field K of characteristic p is _perfect_ if every element of K is a p-th power in K.

THEOREM 48. K _is perfect if and only if every finite-dimensional extension of_ K _is separable over_ K.

Proof. Suppose that the extensions of K are separable. If an element $a \in K$ has no p-th root in K we form $K(u)$ with u a root of $x^p - a$. The irreducible polynomial for u (it is in fact $x^p - a$, as will be proved in the next section) has all its roots equal, so u is not separable, a contradiction.

Conversely, suppose that K is perfect. Let u be algebraic over K and f its irreducible polynomial. If u is not separable, then f is actually a polynomial in x^p. By extracting the p-th root of each coefficient of f we can write f itself as a p-th power, contradicting the irreducibility of f.

Exercises

1. If L is purely inseparable over K and M is purely inseparable over L, then M is purely inseparable over K.

2. The notation is that of Theorem 46 and P is another field between K and N.

(a) N is purely inseparable over P if and only if $P \supset L$,

(b) If N is separable over P, then $P \supset M$,

(c) If $P \cap L = K$, then $P \subset M$.

3. Let f be an irreducible polynomial over K and suppose that (in a splitting field) f has all its roots equal. Show that the characteristic of K must be $p \neq 0$, and f must have the form $x^{p^n} - a$.

4. Prove: the irreducible polynomial for a purely inseparable element has the form $x^{p^n} - a$.

5. Prove: if L is finite-dimensional and purely
inseparable over K, then $[L:K]$ is a power of p.

6. If K is perfect and L is a finite-dimensional extension
of K, then L is perfect. (The converse is also true; see
Exercise 7 of the next section.)

7. If $M = K(u, v)$ where u and v are algebraic over K
and u is separable, then M is a simple extension of K. (Hint:
prove that there are only finitely many intermediate fields. It
can be assumed that $K(u)$ is the maximal separable subfield.
Analyze an intermediate field L by showing that it lies between
its maximal separable subfield L_0 and $L_0(v)$.)

8. Let $x^4 + ax^2 + b$ be irreducible over a field K of
characteristic 2, and assume $a + c^2 b$ is not a square in K for
any $c \in K$. Let $N = K(u)$ where u is a root. Prove that (in the
notation of Theorem 46) $L \cup M \neq N$.

9. Prove: if u is separable over K, then $K(u) = K(u^p)$.

10. Prove: if u is separable over K and v is purely inseparable over K, then $K(u, v) = K(u + v)$. Also, $K(u, v) = K(uv)$
if $u, v \neq 0$.

12. Miscellaneous results on radical extensions

It is possible to give a complete result on the reducibility of an equation of the form $x^n - a$ over an arbitrary field. We begin by showing that the problem reduces to the case where n is a prime power.

THEOREM 49. <u>Let</u> K <u>be any field,</u> a <u>an element in</u> K, m <u>and</u> n <u>relatively prime integers. Then</u> $x^{mn} - a$ <u>is irreducible over</u> K <u>if and only if both</u> $x^m - a$ <u>and</u> $x^n - a$ <u>are irreducible over</u> K.

<u>Proof</u>. If $x^{mn} - a$ is irreducible, so is $x^m - a$ by elementary algebra, for $x^{mn} - a = (x^n)^m - a$.

Conversely, assume $x^m - a$ and $x^n - a$ irreducible over K. Let u be a root of $x^{mn} - a$. Then u^m is a root of $x^n - a$. Hence $[K(u^m):K] = n$ and similarly $[K(u^n):K] = m$. By Theorem 1, $[K(u^n, u^m):K] = mn$. But $K(u^m, u^n) = K(u)$, since $u = u^{rm} u^{sn}$ for integers r, s with $rm + sn = 1$. Hence the degree of u over K is nm, and $x^{mn} - a$ must be irreducible over K.

We proceed to attack the prime-power case. The complete result is given in Theorem 51, with Theorem 50 as a prelude. An intriguing aspect of the investigation is that the prime 2 behaves quite differently from odd primes.

THEOREM 50. <u>Let</u> p <u>be prime,</u> $x^p - a$ <u>irreducible over</u> K <u>and</u> u <u>a root of</u> $x^p - a$. <u>Then:</u>

(1) <u>If</u> p <u>is odd, or if</u> p = 2 <u>and</u> K <u>has characteristic 2,</u> u <u>is not a p-th power in</u> K(u),

(2) <u>If</u> p = 2 <u>and</u> K <u>has characteristic</u> $\neq$ 2, u <u>is a square in</u> K(u) <u>if and only if</u> -4a <u>is a fourth power in</u> K.

Proof. We suppose that $u = w^p$ for w in $K(u)$ and see what conclusion we can reach. The case where K has characteristic p is simplest of all: since w is a polynomial in u and p-th powers are taken termwise we have $w^p \in K$, $u \in K$, a contradiction. We assume that the characteristic is not p.

Adjoin to L a primitive p-th root of unity, say ϵ. The resulting field M is a splitting field of $x^p - a$ over K and is thus normal over K. Any automorphism of M/K sends u into some $\epsilon^i u$ and there is, for every $i = 0, \ldots, p-1$, an automorphism T_i sending u into $\epsilon^i u$. Write $w_i = wT_i$. Then $\epsilon^i u = w_i^p$. The element w is in $K(u)$ but not in K. Its irreducible polynomial (say g) over K necessarily has degree p, and has p distinct roots in M. If w' is any of these roots there is an automorphism S of M/K sending w into w'. If $uS = \epsilon^j u$ we must have $wS = w_j$. Hence the elements $w_o, w_1, \ldots, w_{p-1}$ yield all the roots of g and must be exactly all the roots of g. We conclude that $z = w_o w_1 \cdots w_{p-1} \in K$. We now multiply together the equations $\epsilon^i u = w_i^p$, finding $\eta u^p = \eta a = z^p$, where $\eta = 1 \cdot \epsilon \cdot \epsilon^2 \cdots \epsilon^{p-1}$. If p is odd, $\eta = 1$, and we have the contradiction $a = z^p$. This completes the proof of part (1) of the theorem.

When $p = 2$, $\eta = -1$ and we find only that $-a$ is a square in in K. The completion of the investigation in this case is so elementary that we shall do it again from scratch.

Write $u = w^2$, $w = \alpha + \beta u$ (α and β in K). From $u = (\alpha + \beta u)^2$ we get the equations $\alpha^2 + \beta^2 a = 0$, $2\alpha\beta = 1$. Eliminating β, we find $a = -4\alpha^4$, so that $-4a = 16\alpha^4$ is a fourth power in K. Conversely if $-4a = 16\alpha^4$, we take $\beta = 1/2\alpha$ and we verify $u = (\alpha + \beta u)^2$.

THEOREM 51. <u>Let</u> p <u>be a prime and</u> a <u>an element in</u> K <u>with no p-th root in</u> K. <u>Then</u>

(1) <u>If</u> p <u>is odd</u>, $x^{p^n} - a$ <u>is irreducible over</u> K <u>for any</u> n,

(2) <u>If</u> $p = 2$ <u>and the characteristic is</u> 2, $x^{2^n} - a$ <u>is irreducible over</u> K <u>for any</u> n,

(3) <u>If</u> $p = 2$, $n \geq 2$, <u>and the characteristic is not</u> 2, $x^{2^n} - a$ <u>is irreducible over</u> K <u>if and only if</u> $-4a$ <u>is not a fourth power in</u> K.

<u>Proof</u>. First we show that $x^p - a$ is irreducible over K. Suppose the contrary, and let f be an irreducible factor of $x^p - a$ of degree k $(0 < k < p)$. Let c be the constant term of f. The roots of $x^p - a$ (in some splitting field) all have the form ϵu, where u is one fixed root and $\epsilon^p = 1$. Since $\pm c$ is a product of k of these roots, we have $\pm c = \eta u^k$, $\eta^p = 1$. There exist integers r and s such that $rk + sp = 1$. We have $u = u^{rk} u^{sp} = (\pm c/\eta)^r a^s$. Hence $u\eta^r$ lies in K. Since its p-th power is a, we have the desired contradiction.

We shall next prove simultaneously parts (1) and (2) of the theorem. Let v be a root of $x^{p^n} - a$ and write $u = v^{p^{n-1}}$. We have $u^p = a$, so that $[K(u):K] = p$ (since we have just proved that $x^p - a$ is irreducible over K). If we show that v has degree p^{n-1} over $K(u)$ it will follow that v has degree p^n over K and $x^{p^n} - a$ is irreducible over K. That v has degree p^{n-1} over $K(u)$ will be true by induction on n provided u is not a p-th power in $K(u)$. This is so, in cases (1) and (2), by Theorem 50.

We proceed to part (3) of the theorem. Assume first that $-4a$ is a fourth power in K and write $a = -4\alpha^4$, $y = x^{2^{n-2}}$. Then
$$x^{2^n} - a = y^4 + 4\alpha^4 = (y^2 + 2\alpha y + 2\alpha^2)(y^2 - 2\alpha y + 2\alpha^2).$$
Conversely, suppose $-4a$ is not a fourth power in K. Again, let

v be a root of $x^{2^n} - a$ and $u = v^{2^{n-1}}$. We have $[K(u):K] = 2$ since $u^2 = a$ and we must prove $[K(v):K(u)] = 2^{n-1}$. For $n = 2$ this will be true if u is not a square in $K(u)$, and for $n > 2$ this will be true by induction on n provided u is not a square in $K(u)$ and $-4u$ is not a fourth power in $K(u)$. In the latter case, $-u$ is a square in $K(u)$. So it will suffice to rule out the possibility that either u or $-u$ is a square in $K(u)$. Now these two statements are in fact equivalent ($u \rightarrow -u$ induces an automorphism of $K(u)$ over K) so that part (3) of Theorem 50 applies to complete the proof.

We supplement Theorem 51 with an analogous one on building towers of extensions in the context of polynomials of the type $x^p - x - a$.

THEOREM 52. Suppose that K has characteristic p, $x^p - x - a$ is irreducible over K, and u is a root of $x^p - x - a$. Then $x^p - x - au^{p-1}$ is irreducible over $K(u)$.

Proof. If on the contrary $x^p - x - au^{p-1}$ is reducible, then (Exercise 2, §7) it has a root v in $K(u)$. Say $v = \alpha + \beta u + \ldots + \gamma u^{p-1}$. We have

$$v^p = \alpha^p + \beta^p u^p + \ldots + \gamma^p u^{p(p-1)}$$
$$= \alpha^p + \beta^p(u+a) + \ldots + \gamma^p(u+a)^{p-1} .$$

In

$$v^p = v + au^{p-1} = \alpha + \beta u + \ldots + (\gamma+a)u^{p-1}$$

we equate coefficients of u^{p-1}. The result is $\gamma^p = \gamma + a$. This contradicts the irreducibility of $x^p - x - a$ over K.

By combining these results with earlier techniques we prove two theorems on the existence of extensions of large degree.

THEOREM 53. <u>Let</u> p <u>be an odd prime and suppose that</u> <u>the field</u> K <u>has an extension whose degree is divisible by</u> p. <u>Then for any</u> n, K <u>has an extension whose degree is divisible</u> <u>by</u> p^n. <u>For</u> p = 2 <u>the conclusion is still correct if</u> K <u>has</u> <u>characteristic</u> 2.

<u>Proof.</u> We give a more or less unified proof, but the characteristic of K must receive occasional attention. First suppose the characteristic is p. Then if K is not perfect, we are done; for there will exist an element a in K with no p-th root in K and then Theorem 51 furnishes the irreducible polynomial $x^{p^n} - a$. So if the characteristic is p we may assume that K is perfect.

Let M be the given extension of K with [M:K] divisible by p. Let L be the maximal separable subfield of K. By part (4) of Theorem 46, M is purely inseparable over L. By Exercise 5, §11, [M:L] is a power of the characteristic. If the characteristic is not p, [L:K] is therefore still divisible by p. If the characteristic is p, M = L by Theorem 48 (since we are assuming K perfect) and again [L:K] is divisible by p. In other words, we may assume that the given extension is separable. By passing to a normal closure we may further assume that it is normal. By invoking Galois theory and the existence in the Galois group of an element of order p, we may still further assume that we are dealing with a normal extension of degree p. (In doing this, we have of course replaced K by a larger field, but this does not change the problem.)

Let us then start the notation fresh with L normal over K and [L:K] = p. Again the characteristic makes a difference. If the characteristic is p, then by Theorem 32, L = K(u) with u a root of an irreducible polynomial over K of the form $x^p - x - a$.

Application of Theorem 52 yields an extension of K of degree p^2 and the process can be iterated to get degree p^n for any n. If the characteristic is not p we must pay our respects to the possibly missing p-th roots of 1. Let N be a splitting field of $x^p - 1$ over L, and let M be K with the roots of unity adjoined. A simple argument (of the type used repeatedly in § 6) shows that N is normal over M with Galois group cyclic of order p. By Theorem 34, $N = M(u)$ with u a root of an irreducible polynomial $x^p - a$. It only remains to apply Theorem 51.

THEOREM 54. If K has an extension with degree divisible by 4, then for any $n \geq 2$, K has an extension with degree divisible by 2^n.

Proof. If the characteristic is 2, the result is covered by Theorem 53; so we assume characteristic different from 2. A virtual repetition of the argument in Theorem 53 enables us to reorganize the given extension so that L is normal over K with $[L:K] = 4$. We now further adjoin i with $i^2 = -1$. Then $L(i)$ is normal over $K(i)$ with relative degree 2 or 4. In either case, $L(i)$ contains a quadratic extension of $K(i)$ and hence $K(i)$ contains an element a with no square root in $K(i)$. It cannot be the case that $-4a$ has a fourth root in $K(i)$, for then $-a$ is a square and so is a. By Theorem 52, $K(i)$ therefore has extensions of degree 2^n for any n. This completes the proof of Theorem 54.

There is still one situation for us to examine. If K has an extension of even degree but no extension with degree divisible by 4, then a reorganization as above will lead to a pair $K \subset L$ with $[L:K] = 2$ and no extensions of K with degree divisible by 4. Even the weaker assumption that K has no extensions exactly of degree 4 suffices for a strong conclusion.

THEOREM 55. <u>Let</u> K <u>be a field which has a quadratic ex-</u>
<u>tension but no extension of degree</u> 4. <u>Then</u> K <u>is an ordered field</u>
<u>in which every positive element has a square root.</u>

Proof. Theorem 53 rules out the possibility that K has
characteristic 2. Next, by Theorem 51 it must be the case for
every a in K that either a is a square or -4a is a fourth
power; otherwise x^4 - a would be irreducible over K and yield
an extension of degree 4. In particular, either a or -a is a
square. It cannot be the case that -1 is a square, for then
every element in K would be a square, and no quadratic extension of K could exist. Thus, for any a, either a or -a is a
square but not both. We form the field K(i) with i^2 = -1. In
K(i) every element must be a square, otherwise we would get a
quadratic extension of K(i) and thereby an extension of K of
degree 4. Writing out the fact that a + bi is a square in K(i),
we find that $a^2 + b^2$ is a square in K. This supplies what is
needed to show that K can be ordered by decreeing that the positive elements shall be precisely the squares (up to this point we
would not have known that the sum of positive elements is positive).

A partial converse to Theorem 55 appears as Exercise 5
below.

We have accumulated all that is needed to prove briefly a
pretty theorem of Artin and Schreier.

THEOREM 56. <u>Suppose a field</u> K <u>is not algebraically</u>
<u>closed but has a finite-dimensional extension</u> L <u>which is alge-</u>
<u>braically closed. Then</u> K <u>is an ordered field and</u> L = K(i),
i^2 = -1.

Proof. The hypothesis puts a fixed bound (the degree of L
over K) on the degree of any extension of K. Therefore, by
Theorems 53 and 54, [L:K] must be 2. Then, by Theorem 55,

K is an ordered field in which every positive element has a square root, and L must be K(i).

The last two theorems have of course been reminiscent of the facts which hold when K is the field of real numbers, L the field of complex numbers. But we can add something more to the picture. The theorem that the field of complex numbers is algebraically closed (in ancient days called the "fundamental theorem of algebra") can be given a neat proof by Galois theory. We first prove:

THEOREM 57. Suppose that for some prime p every extension of K has degree divisible by p. Then every extension of K has degree a power of p.

Proof. Let M be an extension of K. We must prove that [M: K] is a power of p. We can assume that M is separable over K. For if the characteristic is a prime different from p, our hypothesis certainly implies that K is perfect; while if the characteristic is p we take the maximal separable subfield of M, noting that M is purely inseparable over it and so this upper degree in any event is a power of p. Furthermore by passing to a normal closure we can (change notation again) assume that M is normal over K. Let P be a p-Sylow subgroup of the Galois group of M/K, and let L be the corresponding subfield. Then [L: K] = the index of P and hence is prime to p. By our hypothesis this is possible only if L = K. Hence [M: K] is a power of p.

We shall now prove that the field of complex numbers is algebraically closed, using a minimum of information from analysis. All we need is: (1) every positive real number has a real square root, (2) every polynomial of odd degree over the real numbers has a real root.

THEOREM 58. <u>Let</u> K <u>be an ordered field in which every</u> <u>positive element has a square root. Suppose further that every</u> <u>polynomial of odd degree over</u> K <u>has a root in</u> K. <u>Then</u> K(i) <u>is algebraically closed, where</u> $i^2 = -1$.

Proof. Our hypothesis implies that K has no extensions of odd degree. By Theorem 57 the degree of any extension of K is therefore a power of 2. Let N be a normal extension of K. The order of the Galois group G of N/K is a power of 2. If [N: K] > 2 then, by group theory, G has a subgroup of index 4 which in turn is contained in a subgroup of index 2. Transferring this information to the intermediate fields we get $K \subset L \subset M$ with [L: K] = [M: L] = 2. Necessarily L is K(i), for this is the only quadratic extension of K. But now Exercise 5 shows that M cannot exist, for every element in L has a square root in L. So: we have proved that the only extension of K (other than K itself) is K(i). Hence K(i) is algebraically closed.

We return to the study of the field K(u) where u is a root of an irreducible polynomial $x^p - a$. In Theorem 59 we show that (for characteristic $\neq p$) K(u) contains no p-th roots of elements of K other than the obvious ones. We use this information to show in Theorem 60 that the adjunction of two "genuinely different" p-th rooths result in an extension of degree p^2.

THEOREM 59. <u>Let</u> p <u>be a prime other than the charac-</u> <u>teristic of</u> K. <u>Let</u> L = K(u) <u>with</u> u <u>a root of an irreducible</u> <u>polynomial over</u> K <u>of the form</u> $x^p - a$. <u>Then: an element</u> v <u>in</u> L <u>satisfies</u> $v^p \in K$ <u>if and only if</u> v <u>has the form</u> bu^n (b $\in$ K).

Proof. If $v = bu^n$ then $v^p = b^p a^k \in K$. Conversely, sup-pose $v \in K(u)$ satisfies $v^p = c \in K$. Say $v = d_0 + d_1 u + \ldots + d_{p-1} u^{p-1}$. If $v \neq 0$ then some $d_i \neq 0$. By multiplying by an appropriate power of u we can switch this d_i to the constant term. In other words we may assume $d_0 \neq 0$. After this normalization we shall prove that v lies in K. Suppose the contrary; then the degree of v over K is necessarily p and $x^p - c$ must be the irreducible polynomial for v over K. Now, just as in the proof of Theorem 51, we argue that the other roots of $x^p - c$ are obtained by replacing u by $\epsilon^i u$ ($i = 1, \ldots, p-1, \epsilon^p = 1, \epsilon \neq 1$) in $d_0 + d_1 u + \ldots + d_{p-1} u^{p-1}$. The sum of all the roots of $x^p - c$ is 0 (for there is no term in x^{p-1}). Moreover for an η satisfying $\eta^p = 1, \eta \neq 1$ we have $1 + \eta + \ldots + \eta^{p-1} = 0$. It follows that in summing the roots all the terms in u drop out and we get simply pd_0. Since $p \neq 0$ and $d_0 \neq 0$ we have the desired contradiction.

THEOREM 60. _Let_ p _be a prime other than the charac-teristic of_ K. _Let_ u, v _be the roots of irreducible polynomials_ $x^p - a$, $x^p - b$ _over_ K. _Then_ $[K(u,v):K] = p^2$ _unless_ $b = c^p a^n$ _for some_ n _and some_ $c \in K$.

Proof. If the polynomial $x^p - b$ remains irreducible over K(u) then $[K(u,v):K] = p^2$. If it factors over K(u) then (Theorem 51) it has a root w in K(u). Necessarily w has the form $w = \epsilon v$ with $\epsilon^p = 1$. By Theorem 59 we have $w = cu^n$ (c $\in$ K). Raising this equation to the p-th power we get $b = c^p a^n$.

As our final topic in this section we shall examine the question of computing the degree of $u + v$ when u and v are given elements algebraic over a field K. We make no attempt to be ex-haustive; but Theorems 63 and 64 do cover some useful territory. Two preliminary theorems will be proved first.

THEOREM 61. Let M be normal and finite-dimensional over K and let u, v be elements of M. Let $u = u_1, u_2, \ldots, u_m$ be the conjugates of u (i.e., the roots of its irreducible equation over K) and let $v = v_1, \ldots, v_n$ be the conjugates of v. Assume finally that $[K(u, v):K] = mn$. Then: for any i, j there exists an automorphism of M/K sending u into u_i and v into v_j.

Proof. We construct instead the inverse automorphism sending u_i into u and v_j into v. There is some automorphism of M/K sending u_i into u. We have as yet no control over what it does to v_j, but after normalizing by applying this automorphism, our problem reduces to constructing an automorphism of M/K keeping u fixed and sending some v_k into v. Now our hypothesis implies that the degree of v over K(u) is still n, so that the roots of its irreducible polynomial over K(u) are still $v_1, \ldots, v_n$. Since M is normal over K(u), the desired automorphism exists.

Now suppose that (in the setup of Theorem 61) we wish to find the degree of u+v over K. The theorem says that all the elements $u_i + v_j$ (i = 1, ..., m, j = 1, ..., n) are conjugates of u+v. If all these mn elements are distinct we conclude that the degree of u+v is mn. If a coincidence occurs, then the difference of two u's will equal the difference of two v's. This is the motivation for studying the difference of two conjugates as we do in the next theorem.

THEOREM 62. Let M be the splitting field over K of a separable irreducible polynomial f having a prime degree p different from the characteristic of K. Let u_1, u_2 be two distinct roots of f. Then M is a normal closure of the field $K(u_1 - u_2)$.

Proof. Let L be the normal closure of $K(u_1 - u_2)$ in M; we have to prove $L = M$. The Galois group of M/K contains an automorphism T of order p which is necessarily a p-cycle on the roots of f. We can suppose the roots numbered so that the cycle is $(u_1 u_2 \ldots u_p)$. Since T sends L into itself we find that $u_1 - u_2,\ u_2 - u_3, \ldots, u_{p-1} - u_p$ are all in L. It follows that $u_1 - u_i \in L$ for every i. Adding up, and recalling that the sum of the u's is even in K, we get $pu_1 \in L$. Since the characteristic is not p, $u_1 \in L$, hence every $u_i \in L$, and $L = M$.

THEOREM 63. Let u, v be non-zero elements separable over K. Suppose the degree of u over K is a prime p different from the characteristic of K, and that the degree of v over K is $n < p$. Then $K(u+v) = K(u, v)$, i.e., $u+v$ has degree pn over K.

Proof. We work in a normal closure N of $K(u, v)$. Let M, L be the normal closures (within N) of $K(u)$ and $K(v)$ respectively. All will be done by Theorem 61 if we show that no difference of two conjugates of u equals a difference of two conjugates of v. Suppose $u_i - u_j = v_r - v_s$; then $u_i - u_j \in L$. But by Theorem 62, this implies $M \subset L$. This is impossible since $[L:K]$ is a divisor of $n!$ and hence is not divisible by p.

THEOREM 64. Let p be a prime different from the characteristic of K. Let u, v be roots of irreducible polynomials of the form $x^p - a$, $x^p - b$, over K. Assume that $[K(u, v):K] = p^2$. Then $u+v$ has degree p^2 over K.

Proof. Again we have only to check that no difference of two conjugates of u equals a difference of two conjugates of v. Now the difference of two conjugates of u has the form $(\epsilon^i - \epsilon^j)u$ where ϵ is a primitive p-th root of unity. So the existence of such an equation would yield the conclusion that u/v lies in $K(\epsilon)$. But

$[K(\epsilon): K] \leq p-1$, while the degree of u/v over K must divide $p^2 = [K(u, v): K]$. Thus $u/v \in K$ would be forced, a contradiction of $[K(u, v): K] = p^2$.

Exercises

1. Let K be a field in which $-1, 2,$ and -2 are not squares. Prove that $x^{2^n} + 1$ is irreducible over K for any n. (Note that this is the cyclotomic polynomial of primitive 2^{n+1}-th roots of 1). State and prove a converse.

Exercises 2 and 3 are multiplicative analogs of Theorems 62 and 63.

2. Let M be a splitting field over K of a separable irreducible polynomial of prime degree p. Let u_1, u_2 be two distinct roots of f. Assume that u_1^p does not lie in K (that is, f, taken with highest coefficient 1, is not of the form $x^p - a$). Prove that M is a normal closure of $K(u_1/u_2)$.

3. Let u, v be elements separable over K. Suppose the degree of u over K is a prime p, that u^p does not lie in K, and that the degree of v over K is $n < p$. Prove that $K(uv) = K(u, v)$ and that uv has degree pn over K.

4. Let u, v be elements separable over K. Suppose the degree of v over K is a prime p and that $u^n \in K$ for some n less than p. Prove that $K(u, v) = K(uv)$ and that uv has degree pn over K. (Hint: use Theorem 61 and the multiplicative plan of attack again. Defeat is possible only if $v_i = \epsilon v_j$ for two conjugates of v and an n-th root of 1. Note that adjunction of ϵ does not change the irreducible equation for v, and examine how that equation changes when v is replaced by ϵv.)

5. Let K be an ordered field in which every positive element is a square. Prove that in $K(i)$, $i^2 = -1$, every element is a square.

6. Find the degrees over the rational numbers of
  (a) $3^{1/5} + 2^{1/4}$,
  (b) $2^{1/3} - 7^{1/3}$,
  (c) $(1 + 3^{1/5})5^{1/4}$,
  (d) $2^{1/5}3^{1/4}$.

7. Let K be a field of characteristic p. Suppose that K has a finite-dimensional extension which is perfect. Prove that K is perfect.

8. What is the degree over the rational numbers of $2^{1/4} + 3^{1/4}$? Can you "embed" this in a more general theorem?

9. Let p be an odd prime different from the characteristic of K. Assume that $(x^{p^2} - 1)/(x^p - 1)$ is irreducible over K. Prove that for any n,
$$(x^{p^n} - 1)/(x^{p^{n-1}} - 1)$$
is irreducible over K.

10. (a) If $x^6 + x^3 + 1$ is irreducible over a field K, prove that $x^2 + x + 1$ and $x^3 - 3x + 1$ are irreducible.
  (b) If $x^2 + x + 1$ and $x^3 - 3x + 1$ are irreducible over K, prove that $(x^{3^n} - 1)/(x^{3^{n-1}} - 1)$ is irreducible over K for any n.

13. Infinite Algebraic Extensions

The theory of infinite-dimensional algebraic extensions presents us with very few surprises. Nearly everything we have proved remains valid, with the exception of results referring specifically to integers occurring as dimensions. The only serious hurdle to be overcome is the existence of suitable extensions, above all the existence of an algebraic closure.

DEFINITION. A field L containing K is said to be an algebraic closure of K if L is algebraically closed and is algebraic over K.

The difficulty in proving the existence of an algebraic closure is set-theoretic rather than algebraic. For the reader who is willing to accept an old fashioned transfinite induction, the following sketch is offered: let $\{f_\lambda\}$ be a well-ordering of the irreducible polynomials over K, for λ a limit ordinal define L_λ to be the union of L_α for $\alpha < \lambda$, and define $L_{\lambda+1}$ to be a splitting field of f_λ over L_λ. (Note: the final touch in this proof is supplied by Exercise 1.)

However, the temper of the times requires that (even at the expense of some gymnastics) we carry through the proof using instead Zorn's lemma. But we must not simply say "Apply Zorn's lemma to the set of all algebraic extensions of K" ; this naive gambit would make anyone familiar with the paradoxes turn pale. We cautiously first get a bound on the cardinal numbers involved.

THEOREM 65. Let L be an algebraic extension of K. If K is infinite, L and K have the same cardinal number. If K is finite, L is either finite or countable.

Proof. We assume K infinite; the argument for K finite is just a slight variant.

Standard set-theoretic arguments show that the number of polynomials over K is the same as the cardinal number (say $\aleph$) of K. With each irreducible polynomial over K associate whatever roots it has in L. This covers all the elements of L with no duplication. This gives us the upper estimate $\aleph_0 \aleph = \aleph$ for the cardinal number of L.

Even with this cardinal number bound established, it would be daring to contemplate all algebraic extensions of K. To treat the matter more cautiously we name a fixed set S disjoint from K; to give ourselves plenty of room we take the cardinal number of S to be greater than that of K (and also greater than $\aleph_0$ if K is finite). The only fields we allow for the discussion have as underlying set K and a subset of S; the field operations on K are to be maintained unchanged. Thus restricted, we are able to apply Zorn's lemma to the set of algebraic extensions of K and pick a maximal field L. L must be algebraically closed; for by Theorem 65 there are still plenty of elements left in S to construct an algebraic extension of L if an algebraic extension were still possible. We have proved:

THEOREM 66. Any field has an algebraic closure.

The uniqueness of algebraic closure might as well be discussed in the more general context of splitting fields.

DEFINITION. Let K be a field, $\{f_\alpha\}$ a set of polynomials with coefficients in K. A field $L \supset K$ is said to be a splitting field of $\{f_\alpha\}$ over K if each f_α factors completely in L and L can be obtained from K by adjoining the roots of the f's.

As a first exercise, the reader should verify that Theorem 25 holds for infinite-dimensional extensions.

THEOREM 67. <u>Let</u> K, K_o <u>be fields and</u> S <u>an isomorphism of</u> K <u>onto</u> K_o. <u>Let</u> $\{f_\alpha\}$ <u>be a set of polynomials with coefficients in</u> K, <u>and let</u> $\{f_{\alpha, o}\}$ <u>be the corresponding polynomials over</u> K_o. <u>Let</u> N <u>be a splitting field of</u> $\{f_\alpha\}$ <u>over</u> K <u>and</u> N_o <u>a splitting field of</u> $\{f_{\alpha, o}\}$ <u>over</u> K_o. <u>Then:</u> S <u>can be extended to an isomorphism of</u> N <u>onto</u> N_o.

Proof. Once again all the algebraic difficulties lie behind us, and only standard set-theoretic maneuvers need to be carried out. To use Zorn's lemma we set up a fussily defined partially ordered set. Its elements are triples (P, P_o, T) where P is a field be-tween K and N, P_o a field between K_o and N_o, and T an iso-morphism of P onto P_o which extends S. We say $(P, P_o, T) \geq (Q, Q_o, U)$ if $P \supset Q$, $P_o \supset Q_o$ and T is an extension of U. All is well for the application of Zorn's lemma and we ob-tain a maximal triple (L, L_o, V). We have to show that $L = N$ and $L_o = N_o$. If for instance $L \neq N$, then some polynomial f_α has not factored completely in L. Define M $(L \subset M \subset N)$ by adjoining the roots of f_α to L; note that M is a splitting field of f_α over L. Similarly define M_o by adjoining the roots of $f_{\alpha, o}$ to L_o. By Theorem 21, V can be extended to an isomorph-ism of M onto M_o. This contradicts the maximality of (L, L_o, V).

In order to apply Theorem 67 to see the uniqueness of alge-braic closure, we have only to note that an algebraic closure is a splitting field -- see Exercise 2.

The major point that remains to be settled is the generaliza-tion of Theorem 24. We will leave the proof to the reader: the fact to be established is that an algebraic extension M of K is normal over K if and only if it is separable over K and a splitting field over K. It is then immediate that M is also normal over any intermediate field L. Hence:

THEOREM 68. Let M be normal and algebraic over K. Then in the Galois correspondence every intermediate field is closed.

As for the subgroups, it is not true that they are closed. The question is considerably clarified by an observation of Krull [Mathematische Annalen (1928), vol. 100]: it is possible to topologize the Galois group G of an algebraic extension in such a way that G becomes a compact topological group, and a subgroup of G is closed in the sense of Galois theory if and only if it is topologically closed. Now the existence of non-closed subgroups can be seen purely group-theoretically, for it is easy to prove that any infinite compact group contains a subgroup which is not closed.

Exercises

1. Let L be an algebraic extension of K with the property that every polynomial with coefficients in K factors completely in L. Prove that L is algebraically closed.

2. Let L be an algebraic closure of K. Prove that L is a splitting field over K of all polynomials with coefficients in K, or of all irreducible polynomials over K.

3. State and prove the appropriate generalization to infinite algebraic extensions of Theorem 46.

PART II. RINGS

Introduction

As a guest of UCLA in the Spring of 1955, I gave a course on ring theory. Notes on most of the course were prepared by Kenneth Hoffman. He wrote the following in a preface:

"These notes represent the essential content of a series of 27 lectures given by Professor Irving Kaplansky at the University of California, Los Angeles, during the spring semester, 1955. An effort has been made to make the notes self-contained; however certain elementary definitions such as ring, ideal, etc. have been omitted to save space. The basic subject matter is related in a sequence of theorems, with numbers running consecutively through the chapters. Theorems dealing with material which is not pure ring theory are generally indexed with letters or the unqualified word "theorem". Much information is presented under the general heading "Remarks". The remarks made are not guaranteed to be obvious (although some are).

"The final chapter represents a considerable extension of the subject of simple rings as presented in the lectures. This chapter was written directly by Professor Kaplansky. "

After a subterranean existence for nearly ten years, the notes surfaced in February, 1965 in the Chicago Mathematics Department's lecture notes series. They were unchanged except for the addition of a preface summarizing recent developments.

In the present reprinting there has been some editing, and at appropriate places new material has been added.

This account of ring theory should be regarded as a supplement to the literature. The reader should have on hand the four

great classics from the days of chain conditions: Deuring's
Algebren, Albert's Structure of Algebras, Rings with Minimum
Condition by Artin, Nesbitt and Thrall, and Jacobson's Theory of
Rings, as well as Bourbaki's Modules et Anneaux Semi-simples,
Herstein's Carus Monograph on Non-commutative Rings, and of
course the standard work: Jacobson's Structure of Rings.

I am very grateful to Kenneth Hoffman for his excellent job
on the notes, and for allowing me to incorporate his work into
Fields and Rings.

1. The Radical

DEFINITION. Let A be a ring. A right A-module is an abelian group M, over which there is defined an external law of composition $(x, \alpha) \rightarrow x\alpha$, with elements of A, in such a way that for all x, y in M and α, β in A:

(1) $(x + y)\alpha = x\alpha + y\alpha$,

(2) $x(\alpha + \beta) = x\alpha + x\beta$,

(3) $x(\alpha\beta) = (x\alpha)\beta$.

A left A-module is a system M, where (3) is replaced by:

(3)' $x(\alpha\beta) = (x\beta)\alpha$.

For left A-modules, it is customary to write the elements of A on the left, so that (3)' assumes the more natural form

(3)' $(\alpha\beta)x = \alpha(\beta x)$.

If M is a right A-module and A has a unit 1 such that $x \cdot 1 = x$, for all x in M, then M is called unitary.

DEFINITION. If M is a right A-module and x is in M, the annihilator of x is the set of elements α in A for which $x\alpha = 0$. If S is a subset of M, the annihilator of S is the intersection of the annihilators of the elements of S.

THEOREM 1. Let A be a ring, and let M be a right A-module. Then the annihilator of a subset S of M is a right ideal in A. If S is a submodule, the annihilator is a 2-sided ideal.

Proof. Let α and β annihilate S. For any x in S, $x(\alpha - \beta) = x\alpha - x\beta = 0 - 0 = 0$. Let γ be any element of A. Then, $x(\alpha\gamma) = (x\alpha)\gamma = 0 \cdot \gamma = 0$. Thus the annihilator is a right ideal. If S is a submodule, then for any γ in A and any α annihilating S we have: $x(\gamma\alpha) = (x\gamma)\alpha = 0$, since $x\gamma$ lies in S. Thus $\gamma\alpha$ annihilates S, and the annihilator is now a 2-sided ideal.

Remark: If I is the annihilator of M, then M is also a right A/I-module.

DEFINITION. The right A-module M is faithful, if the annihilator of M consists exactly of the zero element of A.

DEFINITION. We say that the right A-module M is irreducible, if it has no proper submodules and is not a trivial module, i.e., $M \cdot A \neq 0$.

DEFINITION. A ring is right primitive, if it admits a faithful irreducible right module. (Left primitive ring is defined similarly.)

Remarks: 1. A trivial ring (all products zero) admits no irreducible module.

2. G. M. Bergman (Proc. Amer. Math. Soc. 15(1964), 473-5; correction on page 1000) has given an example of a right primitive ring which is not left primitive.

THEOREM 2. Let x be an element of the right A-module M, and let I be the annihilator of x. Then xA is isomorphic to the module A/I.

Proof. (Note that if I is a right ideal in A, we cannot induce a ring structure on the cosets of I, unless I is also a left ideal; however, we can give these cosets the structure of a right A-module, in the obvious manner.) We define a mapping of A onto xA by $h(\alpha) = x\alpha$. Now h is a module homomorphism. The kernel of h is obviously I, and as usual for algebraic systems, we must have xA isomorphic to A/I.

THEOREM 3. If M is an irreducible right A-module and x is a non-zero element of M, then xA = M.

Proof. Certainly xA is a submodule of M. Thus, either xA = 0 or xA = M; we wish to exclude the former possibility. Let S be the set of elements y in M for which yA = 0. Clearly S is a submodule of M, and since M is non-trivial, S is not all of M. Therefore, S = 0, and in particular, x is not in S. It follows that xA = M.

DEFINITION. The right ideal I in the ring A is called regular if there is an element e in A for which ea - a lies in I for every a in A. The element e is called a left unit modulo I. We note that if e is in I, then I = A.

Example. In the ring of even integers, the ideal (6) is regular (e = 4). The ideal (4) is not regular.

THEOREM 4. If M is an irreducible right A-module, and x is a non-zero element of M, then the annihilator I of x is a regular maximal right ideal in A.

Proof. We know from Theorem 1 that I is a right ideal. We must show two things:

(a) I is regular. By Theorem 3, xA = M; in particular, there is an element e in A such that xe = x. Let a be an element of A. Then x(ea - a) = xea - xa = xa - xa = 0; or, ea - a belongs to I.

(b) I is maximal. If J were a right ideal properly between A and I, then J/I would be a module properly between A/I and (0). But A/I ≅ xA = M, and M has no proper submodule.

THEOREM 5. If M is an irreducible right A-module, then there is a regular maximal right ideal I in A such that M is module isomorphic to A/I.

Proof. See Theorems 2, 3 and 4.

DEFINITION. The 2-sided ideal P in A is called a right primitive ideal if A/P is a right primitive ring.

Remark. If M is a right A-module, then M is a faithful right A/P- module, where P is the annihilator of M. Thus, a 2-sided ideal P is right primitive if and only if P is the annihilator of an irreducible right A-module.

THEOREM 6. A right primitive ideal P is the intersection of the regular maximal right ideals containing it.

Proof. P is the annihilator of an irreducible right A-module M. If x is a non-zero element of M, then the annihilator of x is, by Theorem 4, a regular maximal right ideal in A. The annihilator of M is, by definition, the intersection of the annihilators of elements x in M. The statement of the theorem is now obvious.

THEOREM 7. The intersection of the right primitive ideals of a ring A is the intersection of the regular maximal right ideals of A.

Proof. According to Theorem 6, it will suffice to show that every regular maximal right ideal I contains a right primitive ideal. Now A/I is an irreducible right A-module (non-trivial, since I is regular). The annihilator P of A/I is a right primitive ideal. Furthermore, P is contained in I; for let p be an element of P and let e be the left unit for I. Then ep - p lies in I. But clearly ep belongs to I; hence, p is in I.

DEFINITION. An element x in a ring A is called right quasi-regular, if there is a y in A such that x + y + xy = 0. We shall at times abbreviate right quasi-regular to R.Q.R. If A has a unit, x is R.Q.R. if and only if $(1+x)$ is right regular. For convenience, we introduce the notation

$$x \circ y = x + y + xy .$$

The operation $(\circ)$ is associative and has zero as a unit.

DEFINITION. We say that x is *quasi-regular* if x has both a right and left quasi-inverse. Note: because of associativity, these two inverses are necessarily equal. For if $x \circ y = 0$ and $z \circ x = 0$, then $z = z \circ (x \circ y) = (z \circ x) \circ y = y$.

Remark. If A has a unit, then the mapping $f(x) = 1 + x$ is an isomorphism of A under the circle operation with A under ring multiplication, that is, $f(x \circ y) = f(x) f(y)$. Note also that the quasi-regular elements of a ring form a group under the circle operation.

DEFINITION. A *radical ring* is a ring in which every element is quasi-regular.

Examples of Radical Rings.

1. Any trivial ring.

2. Any *nil* ring, i.e., one in which every element is nilpotent. Indeed, in any ring, a nilpotent element is quasi-regular: if $x^n = 0$, then $(-x) \circ (x + \ldots + x^{n-1}) = 0$.

3. Let F be a field and form all formal power series over F with no constant term: $u = a_1 x + a_2 x^2 + \ldots$. Here we have $(1 + u)^{-1} = 1 - u + u^2 + \ldots$. The right side makes sense since each power of x occurs only a finite number of times. From this relation one can easily see how to determine a quasi-inverse for u.

THEOREM 8. *If a right ideal* I *consists entirely of right quasi-regular elements, then* I *as a subring is a radical ring.*

Proof. Let x be an element of I. We must show that x has a (2-sided) quasi-inverse in I. There is a y in A such that $x + y + xy = 0$. Since $y = -x - xy$, y belongs to I. Hence, y has a right quasi-inverse z which is also in I. But now y is quasi-regular and $x = z$. Therefore y is also a left quasi-inverse of x.

DEFINITION. An ideal (left, right, or 2-sided) consisting entirely of quasi-regular elements will be called a radical ideal (appropriately qualified).

THEOREM 9. If I is a radical right ideal and M is a regular maximal right ideal, then I is contained in M.

Proof. Let x be an element of I, and suppose x does not belong to M. Then $I + M$ is the whole ring; in particular, there is an i in I and an m in M such that $i + m = e$, the relative unit of M. By hypothesis, there is an element j in I for which $-i + j - ij = 0$. Since $ij + mj = ej$, we have $-i + j + mj = ej$. Therefore $i = mj - (ej - j)$, so that i is in M. But then e is in M, which is impossible. Thus, I is contained in M.

THEOREM 10. Any proper regular right ideal can be extended to a regular maximal right ideal (with the same relative unit).

Proof. Zorn's lemma.

THEOREM 11. The intersection, N, of the regular maximal right ideals of A is a radical ideal.

Proof. Let x be in N; we will show that x is right quasi-regular. If x is not R.Q.R., then the ideal I consisting of all

elements $xy + y$, for y in A, is a proper regular right ideal
(it does not contain x), with left unit $(-x)$. By Theorem 10, I
can be extended to a regular maximal right ideal M. But then
x is not in M, a contradiction. Therefore, x is R.Q.R., and
by Theorem 8, N is a radical ideal.

THEOREM 12. <u>The following four ideals are identical:</u>

(1) <u>The intersection of the regular maximal right</u>
<u>ideals.</u>

(2) <u>The intersection of the regular maximal left</u>
<u>ideals.</u>

(3) <u>The intersection of the right primitive ideals.</u>

(4) <u>The intersection of the left primitive ideals.</u>

<u>Proof.</u> By Theorem 7, (1) and (3) are identical, call this
ideal N. By symmetry, (2) and (4) are equal, call this ideal N'.
By Theorem 11, N is a radical ideal, which, by the "dual"
theorem of Theorem 9, lies in every regular maximal left ideal,
and hence in N'. Similarly, N' is contained in N, so that
$N' = N$.

<u>Remark.</u> The ideal $N = N'$ is a radical ideal which contains
every other radical ideal; hence, we call N the <u>radical</u> of A.
The radical also contains every left, right, or 2-sided nil ideal.

DEFINITION. The ring A is said to be <u>semi-simple</u>, if
the radical of A consists of the zero element alone.

DEFINITION. A ring A satisfies the <u>descending chain con-</u>
<u>dition</u> (D.C.C.) <u>on right ideals</u>, if every properly descending
chain of right ideals is finite.

DEFINITION. An <u>algebra</u> is a vector space which is also
a ring, in such a way that for all ring elements x, y and scalars α:

$$\alpha(xy) = (\alpha x)y = x(\alpha y).$$

Remark. If an algebra has a unit, any right (or left) ideal is automatically a subspace. If such an algebra is finite-dimensional, the descending chain condition on right (or left) ideals is satisfied.

DEFINITION. Let I be a subset of A and J a right ideal. The symbol IJ will denote the right ideal spanned by the set of products ij, with i in I and j in J. (It is the set of finite sums of such products.)

DEFINITION. The ideal I is _nilpotent_, if for some positive integer n, $I^n = 0$.

LEMMA. _If_ $ax = a$ _and_ $-x$ _is right quasi-regular, then_ $a = 0$.

Proof. If $(-x) \circ y = 0$, $a(-x \circ y) = -a = 0$.

THEOREM 13. _If the ring_ A _satisfies the descending chain condition on right ideals, the radical of_ A _is nilpotent._

Proof. Let N be the radical of A. Then $N \supset N^2 \supset N^3 \supset \ldots$. By hypothesis $N^k = N^{k+1}$, for some k. Let $P = N^k$; we shall show that $P = (0)$. (Note that $P^2 = P$.) Suppose $P \neq (0)$. Among all right ideals I such that $IP \neq (0)$, pick a minimal one I_o (descending chain condition). Then there is an x in I_o such that $xP \neq (0)$. Consequently $(xP)P = xP^2 = xP \neq (0)$; so, we must have $xP = I_o$. Choose an element a in P such that $xa = x$. Since, in particular, a is in N, $-a$ is right quasi-regular. Hence, by the above lemma, $x = 0$, a contradiction. Therefore $P = N^k = (0)$.

DEFINITION. An algebra is _algebraic,_ if every element satisfies a non-trivial polynomial equation.

Remarks. 1. The direct sum of an infinite number of finite-dimensional algebras is algebraic.

2. The algebra of countably infinite matrices over a field, with only a finite number of non-zero entries, is an infinite-dimensional algebraic algebra which is simple.

THEOREM 14. The radical of an algebraic algebra is nil.

Proof. Let x be in the radical, $x^n + \ldots + \alpha x^k = 0$, where $\alpha \neq 0$. Solve for x^k, obtaining $x^k = x^k(\beta_1 x + \ldots - \alpha^{-1} x^{n-k})$. Now $\beta_1 x + \ldots$ lies in the radical. By the lemma prior to Theorem 13, $x^k = 0$, i.e., x is nilpotent.

DEFINITION. In a topological ring, the element x is called topologically nilpotent, if x^n approaches 0 as $n \to \infty$. An ideal is topologically nil, if it consists of topologically nilpotent elements.

Remark. In a compact topological ring, the radical is topologically nil.

THEOREM A. The radical of a Banach algebra A is topologically nil. Any left, right, or 2-sided topologically nil ideal is in the radical.

Proof. For convenience, assume a unit. Let x be in the radical. For all complex λ, λx is in the radical. Thus $(1 - \lambda x)^{-1}$ exists for all λ; furthermore, the function $F(\lambda) = (1 - \lambda x)^{-1}$ is entire. For sufficiently small λ,

$$(1 - \lambda x)^{-1} = 1 + \lambda x + \lambda^2 x^2 + \ldots \ .$$

Hence, this relation holds for all λ, in particular for $\lambda = 1$. Therefore, x^n approaches 0.

Let I be a topologically nil ideal in A, and let x be in I. Then $2x$ is in I, so that $(2x)^n$ approaches 0. Thus, for some K,

$\|x^n\| \leq K2^{-n}$. It follows that $1 + x + x^2 + \ldots$ converges (to $(1-x)^{-1}$) and thus that $-x$ is quasi-regular. We see that I lies in the radical.

2. Primitive Rings and the Density Theorem

DEFINITION. The ring A is simple, if the only two-sided ideals in A are A and (0), and if, in addition $A^2 \neq (0)$.

Remarks. 1. A simple ring is either a radical ring or a primitive ring.

2. For many years the existence of a simple radical ring was an intriguing open question. Then E. Sasiada (Bull. Acad. Polon. Sci. 9(1961), 257) announced the existence of such a ring. A definitive account has appeared in a joint paper of Sasiada and P. M. Cohn (J. of Alg. 5(1967), 373-7).

3. The Sasiada example is however not a nil ring and it remains an open question whether a simple nil ring exists.

4. If a simple ring has a maximal one-sided ideal (not necessarily regular), then it is not a radical ring.

5. It is at times convenient to consider maximal right ideals as being of one of three types:

(a) Those which contain A^2.

(b) Those which do not contain A^2, but are not regular.

(c) Those which are regular.

There exists a ring in which all three types occur; and there exists a simple ring in which types (b) and (c) occur.

Examples of simple rings.

1. Any division ring is simple, having in fact no proper right or left ideals. What one might term a converse of this statement is also true, namely: If A is a ring with no proper right ideals, then A is trivial (with a prime number of elements), or A is a division ring.

2. The ring A of all n by n matrices over a division ring D is simple.

Proof. Certainly $A^2 \neq (0)$. We let e_{ij} denote a matrix unit. A general element of A is a sum $\Sigma \, \alpha_{ij} e_{ij}$, α_{ij} in D, with multiplication performed according to the rule $e_{ij} e_{kl} = e_{il} \delta_{jk}$ (δ_{jk} = the Kronecker delta). Let I be a two-sided ideal in A with a non-zero element $a = \Sigma \, a_{ij} e_{ij}$. Then the element $e_{ii}(a)e_{jj} = \alpha_{ij} e_{ij}$ lies in I, and accordingly e_{ij} is in I. But then for $k = 1, \ldots, n$, $e_{ki} e_{ij} e_{jk} = e_{kk}$ lies in I; hence the sum of the e_{kk}, which is the unit, lies in I, and I is not proper.

If the descending chain condition on right ideals is assumed, there are no further simple rings, according to the theorem of Wedderburn-Artin which we shall prove shortly. But otherwise there are other examples such as the peculiar one we next describe.

3. Let F be a field of characteristic zero, and let A be the ring of "differential polynomials" $\Sigma \, \alpha_{ij} x^i D^j$, with α_{ij} in F, added in the obvious manner, and multiplied according to the rule: $Dx = xD + 1$. (One may think of the elements of A as operators on real valued functions on the real line, with x corresponding to "multiplication by x" and D to differentiation.) The ring A (which is in fact an infinite dimensional algebra) is a principal ideal ring which has no divisors of zero and is simple. For let I be a two-ideal in A with a non-zero element $a = \Sigma \, \alpha_{ij} x^i D^j$. One easily establishes the formulae:

$$Dx^k = x^k D + kx^{k-1}$$
$$D^k x = xD^k + kD^{k-1} \ .$$

Now the element $b = xa - ax$ lies in I. Either b is a non-zero element of I which is of lower degree in x than a is, or a is free of D. If the former situation prevails, we similarly investigate the element xb - bx in I. After a finite number of such investigations we must thus arrive at a non-zero element of I which is free of x or of D. Let us say, for instance, that we obtain a non-zero c in I which is free of D. Then, if c is not a "constant" (constant polynomial), $c' = cD - Dc$ is a non-zero element of I having lower degree in x than does c. Clearly, in a finite number of steps, we arrive at a non-zero "constant" in I. A similar argument holds when c is free of x. In any case, I contains a non-zero "constant", and is not proper.

Example of a primitive, non-simple ring: Let V be an infinite-dimensional left vector space over a division ring D. Let A be the ring of all linear transformations of V into V. Then A has the following properties:

(a) A is not simple. The set F, of all elements of A having finite-dimensional range, is a proper two-ideal.

(b) The ring A is right primitive, since V is a faithful irreducible right A-module.

(c) The ideal F, of linear transformations with finite-dimensional range, is a simple ring: Let I be a two-ideal in F, $I \neq 0$. We will show that I contains every linear transformation U on V which has one-dimensional range. This will suffice to show that $I = F$, since any transformation in F is a finite sum of such transformations U. Let S be in I, with $x_1 S = y_1 \neq 0$. If $xU = y \neq 0$, we can find a B in F such that $xB = x_1$ and $zU = 0$ implies $zB = 0$; then we can find a C in F with one-dimensional range such that $y_1 C = y$. Clearly $BSC = U$; hence, U is in I.

(d) If I is any non-zero two-ideal in A, F is contained in I, by the same proof as in part (c).

(e) V is a faithful irreducible right F-module.

(f) If V has countable dimension, then A/F is simple. We can show this by proving that any two-ideal I which contains F properly is all of A. If I contains an element S with infinite-dimensional range, then we can find (in a manner similar to that of part (c)) elements B and C in A such that BSC = 1 (we omit the details); hence, I is not proper.

DEFINITION. Let M be a right A-module, and let α be a group endomorphism of M (written on the left). We say that α is an A-<u>endomorphism</u> of M if for every a in A and m in M it is true that $\alpha(ma) = (\alpha m)a$. Under the operations

$$(\alpha + \beta)m = \alpha m + \beta m$$
$$(\alpha\beta)m = \alpha(\beta m)$$

the set of A-endomorphisms of M forms a ring.

Examples:

1. If A is the ring of integers, every group endomorphism of M is an A-endomorphism.

2. If M is a vector space over the field A, the A-endomorphisms of M are the linear transformations of M.

THEOREM 15. (Schur's lemma) <u>The ring</u> D <u>of</u> A-<u>endomorphisms of an irreducible module</u> M <u>is a division ring.</u>

<u>Proof</u>. Since D has a unit, we need only show that each non-zero element α in D has an inverse. If $\alpha \neq 0$, then αM is a non-zero submodule of M, which must be M (irreducibility). Let N be the set of elements m in M for which $\alpha m = 0$. Then N is a submodule of M, which is certainly not all of M; hence N = (0). Thus α is a one-one map of M onto M, and has a set-theoretic inverse, α^{-1}. We need only verify that α^{-1} is an A-endomorphism. But

$$\alpha^{-1}(m+n) = \alpha^{-1}(\alpha m' + \alpha n') = \alpha^{-1}(\alpha(m'+n')) = m' + n',$$

and

$$m' + n' = \alpha^{-1}m + \alpha^{-1}n.$$

Similarly,

$$\alpha^{-1}(ma) = \alpha^{-1}((\alpha m')a) = \alpha^{-1}(\alpha(m'a)) = m'a = (\alpha^{-1}m)a.$$

This suffices to show that D is a division ring. We shall refer to D as the commuting division ring.

Remark. If A is a right primitive ring, with faithful irreducible right A-module M, then Schur's lemma shows us that A is isomorphic to a subring of the ring of linear transformations on M, when M is interpreted as a left vector space over the commuting division ring D. In general, A is not the full ring of linear transformations. (We have seen that this is the case when A is the ring of linear transformations of finite-dimensional range over an infinite-dimensional left vector space.) We are thus led to the following concept.

DEFINITION. A set $\mathcal{S}$ of linear transformations on a vector space V is called n-transitive if for any two sets of vectors in V: $x_1, \ldots, x_n$ and $y_1, \ldots, y_n$, with the x's linearly independent, there is a transformation S in $\mathcal{S}$ such that $x_i S = y_i$, i = 1, ..., n. If $\mathcal{S}$ is n-transitive for every positive integer n, then S is called dense (in the ring of all linear transformations on V).

THEOREM 16. (Density theorem) Let A be a right primitive ring, and let M be a faithful irreducible right A-module. Make M into a left vector space over the commuting division ring D (by Theorem 15). Then A is isomorphic to a dense ring of linear transformations on M.

Proof. If x is any non-zero element of M, then $xA = M$ (Theorem 3). Thus A is 1-transitive. Now to prove density it will suffice to show that for any finite-dimensional subspace E of M and any non-zero vector x which is not in M there is a linear transformation in A which annihilates E but not x. The proof will proceed by induction on the dimension of E. The result has been established when E is zero-dimensional; we assume it has been demonstrated for dimension $n-1$. Then, for any n-dimensional subspace E (and a non-zero x not in E) we write $E = F + Dy$, where F is of dimension $n-1$. Let J be the annihilator of F. Then J is a right ideal in A, and yJ is a submodule of M. By the induction hypothesis, $yJ \neq 0$; hence, $yJ = M$. Now suppose that any element in A which annihilates E also annihilates x. Then we can define a mapping α of M into M by $\alpha(yj) = xj$, for each j in J. (α is well-defined, since if $yj_1 = yj_2$ then $(j_1 - j_2)$ annihilates both F and y, hence E, implying that $xj_1 = xj_2$.) One can easily verify that α is an A-endomorphism, that is, α is in D. By definition $(\alpha y - x)J = 0$, and the induction hypothesis thus implies that $\alpha y - x$ is in F. It follows that x is an element of $F + Dy = E$, contrary to assumption. Consequently, there must exist an element of A which annihilates E but not x.

THEOREM 17: If a right primitive ring satisfies the descending chain condition on right ideals, it is the full ring of linear transformations on a finite-dimensional vector space over a division ring. In particular (Wedderburn-Artin), this is true of a simple ring with descending chain condition on right ideals.

Proof. In the notation of Theorem 16, we must show that M is finite-dimensional. Suppose the contrary, and let $x_1, x_2, \dots$

be an infinite linearly independent set in M. Let I_r be the annihilator of $(x_1, x_2, \ldots, x_r)$. Then $\{I_r\}$ is a descending chain of right ideals of A. By the density theorem, $I_r \neq I_{r+1}$. This violates the descending chain condition.

The result of Wedderburn-Artin follows when one realizes that a simple ring with D.C.C. cannot be a radical ring. (A radical ring with D.C.C. is nilpotent, hence, not simple.)

Remark. A particular consequence of Theorem 17 is that a right primitive ring with D.C.C. on right ideals has a unit.

We mention two applications of the density theorem which arose in various contexts.

1. Let A be a right primitive ring in which the square of every element is right quasi-regular. Then A is a division ring (with no square root of -1).

Proof. We know that A is a dense ring of linear transformations on a vector space V. We shall show that V is one-dimensional. If not, take x and y to be linearly independent vectors in V, and find a linear transformation a in A such that $xa = y$ and $ya = x$. Then $xa^2 = -x$ and $ya^2 = -y$. But then for any b in A, $x(a^2 + b + a^2 b) = -x$ and $y(a^2 + b + a^2 b) = -y$, contradicting the hypothesis that a^2 is right quasi-regular. We remark in connection with this result that a semi-simple Banach algebra in which every x^2 is R.Q.R is automatically commutative.

2. If A is a right primitive ring such that for any a, b in A: $a(ab - ba) = (ab - ba)a$, then A is a division ring.

Proof. If x, y are linearly independent vectors in V, one can easily verify that the transformations a (defined in example 1 above) and b, where $xb = y$, $yb = x$, violate the commuting condition of the hypothesis.

We point out in passing a somewhat deeper result: If $c(ab - ba) = (ab - ba)c$ for all a, b, c, then A is actually a field.

THEOREM 16'. (Classical density theorem) Let V be a finite-dimensional vector space over an algebraically closed field F. Let A be an irreducible algebra of linear transformations on V. Then A is the full algebra of linear transformations on V.

Proof. The proof will utilize Theorem 16. First we note that V is actually a faithful irreducible right A-module. This can be seen as follows: If x is a non-zero element of V, then xA is a submodule of V; in fact, since $\lambda(xa) = x(\lambda a)$, we see that xA is a subspace of V. Since V is irreducible (as a vector space) we must have $xA = V$. It follows that if a submodule of V contains a non-zero vector x, it contains $xA = V$, ie., V is an irreducible right A-module. Certainly V is faithful. Next we show that the commuting division ring D (guaranteed by Schur's lemma) is actually an algebra over F. Let α be in D, λ in F, x in V. For any non-zero y in V there is an a in A such that $x = ya$. Thus $\alpha(\lambda x) = \alpha(\lambda \cdot ya) = \alpha(y \cdot \lambda a) = a(y)(\lambda a)$ $= \lambda \cdot \alpha(ya) = \lambda \cdot \alpha(x)$. Therefore D is actually an algebra of linear transformations over the vector space V, and as such is necessarily finite-dimensional. We now quote the theorem that the only finite-dimensional division algebra over an algebraically closed field is the field of scalars; in other words, $D = F$. We may apply Theorem 16 to conclude that A is a dense ring of linear transformations over the vector space V (with F as scalar field). Since V is finite-dimensional, A is the full algebra of linear transformations on V.

Remarks. 1. Theorem 16' could also have been proved by proving propositions analogous to Theorems 15 and 16 for algebras of linear transformations, rather than for rings of endomorphisms.

2. If we had chosen to develop the theory of rings with operators, establishing our present propositions in the presence of operator domains, Theorem 16' would follow immediately; however, it seems inadvisable to treat rings with operators in an introductory course.

3. The hypothesis that F be algebraically closed is essential. For example let F be the field of real numbers and V a two-dimensional space over F. then the algebra A of transformations on V matricially represented by

$$\begin{pmatrix} a & b \\ -b & a \end{pmatrix}$$

is irreducible.

THEOREM 16″. (Burnside's theorem) Let S be a multiplicative semi-group of linear transformations on an n-dimensional vector space V over an algebraically closed field. Suppose V is irreducible under S . Then S contains n^2 linearly independent transformations.

Proof. Let A be the algebra spanned by S . Apply Theorem 16' to A.

THEOREM B. Let S be an irreducible multiplicative semi-group of linear transformations on an n-dimensional vector space V over an algebraically closed field. Suppose that exactly k distinct traces occur in the elements of S . Then S has at most k^{n^2} elements.

Proof. On the algebra of all linear transformations on V we introduce the inner product $(A, B) = \text{Trace}(AB)$, easily seen to be non-singular. Let $c_1, \ldots, c_k$ be the distinct traces that occur. Let A_i ($i = 1, \ldots, n^2$) be n^2 linearly independent elements in S (Theorem 16"). Each X in S satisfies equations $\text{Tr}(A_i X) = b_i$ where $b_1, \ldots, b_{n^2}$ are chosen from the c's. These equations determine X uniquely, so there are at most k^{n^2} choices for X. (This streamlining of Burnside's argument is due to C. Procesi).

DEFINITION. A matrix is unipotent if all its characteristic roots are 1. An equivalent statement is that the matrix has the form identity plus nilpotent.

THEOREM C. (Kolchin) <u>Let</u> S <u>be a multiplicative semi-group of unipotent matrices. Then the elements of</u> S <u>can be simultaneously put in triangular form.</u>

Proof. Let n be the size of the matrices. We argue by induction on n. The case $n = 1$ is trivial.

Case I. The scalar field is algebraically closed. If S is irreducible, then by Theorem B, S has only one element; whereas one matrix is always reducible (here $n > 1$). Hence S is reducible. Then by choosing a basis for the invariant subspace of S and extending it to a complete basis, all the elements of will have matrices of the block form:

$$\begin{pmatrix} B & C \\ O & D \end{pmatrix}.$$

Now the sets S_L, of the upper left corners B, and S_R, of the lower right corners D, form multiplicative semi-groups of unipotent matrices of dimension less than n. One can then use the induction hypothesis to triangulate simultaneously these matrices, and all elements of S will then have been put in triangular form.

Case II. An arbitrary scalar field F. Form the algebraic closure of F and triangulate the elements of S simultaneously as matrices over the extension field. Then any product of n matrices $(T - I)$, where T is in S and I is the identity matrix, must be zero. Let r be the smallest integer such that the product of any r elements $(T - I)$ is zero. Then there exist elements $T_1, \ldots, T_{r-1}$ in S such that

$$(T_1 - I)(T_2 - I) \cdots (T_{r-1} - I) \neq 0.$$

Find a vector x such that

$$x(T_1 - I) \cdots (T_{r-1} - I) = y \neq 0.$$

Then for any T in S , $y(T - I) = 0$, or $yT = y$. This shows that S is reducible. The argument can now proceed as in Case I.

DEFINITION. A <u>torsion group</u> is a group in which every element is of finite order.

DEFINITION. A group is <u>locally finite</u> if every finitely generated subgroup of it is finite.

The following problem, which comes in two varieties, was proposed by William A. Burnside.

<u>Burnside's Strong Problem</u>: Is every torsion group locally finite?

<u>Burnside's Weak Problem</u>: Is every torsion group, in which the order of every element is less than a fixed integer, locally finite?

We make an analogous definition for algebras.

DEFINITION. An algebra is <u>locally finite</u> if every finitely generated subalgebra is finite-dimensional.

Kurosch proposed the analogues of Burnside's problems.

<u>Kurosch's Strong Problem</u>: Is every algebraic algebra locally finite?

<u>Kurosch's Weak Problem</u>: Is an algebraic algebra locally finite if every element satisfies a polynomial of degree less than a fixed integer?

Kurosch's weak problem was answered in the affirmative by Jacobson for semi-simple algebras, and by Levitzki for radical (i.e., nil) algebras. It is easy to put these two results together to get that the answer to the weak Kurosch problem is "yes".

In a brilliant piece of work Golod (Izv. Akad. Nauk SSSR 28(1964), 273-6), basing himself on ideas of Golod and Shafarevich (same Izv. 261-272) got a negative example for both strong problems. This negative result increases the interest in special cases where the answer is affirmative, such as algebras with a polynomial identity (see section 7).

The status of the weak Burnside problem is uncertain. Novikov (Dokl. Akad. Nauk SSSR 127(1959), 749-752) has announced a negative answer when the bound is at least 72, but full details are not available (see the review by Bruck (Math. Rev. 21, 1051-2)).

For the important class of groups faithfully represented by matrices the answer to the strong Burnside problem (let alone the weak one) is "yes". We proceed to prove this.

THEOREM D. (O. Schmidt) If a group G has a normal subgroup H such that both H and G/H are locally finite, then G is locally finite.

Proof. Let $a_1, \ldots, a_r$ be elements of G, and let $a_1', \ldots, a_r'$ be their respective images under the canonical homomorphism of G onto G/H. These images generate a finite subgroup of G/H with elements $a_1', \ldots, a_r', a_{r+1}', \ldots, a_n'$. Choose any elements $a_{r+1}, \ldots, a_n$ in G which map respectively onto $a_{r+1}', \ldots, a_n'$. For each i, j we have $a_i a_j = h_{ij} a_k$, for some k and some element h_{ij} in H. The elements h_{ij} generate a finite subgroup T of H. Now $a_i a_j a_m = h_{ij} h_{km} a_\ell$ (for some ℓ). Thus the product of any number of the elements a_i is equal to an element of T times some a_ℓ. Clearly then the a_i generate a group with fewer elements than $n \cdot (\text{order } T)$.

THEOREM E. A solvable torsion group G is locally finite.

Proof. We have $G = G_o \supset G_1 \supset \cdots \supset G_n = 1$, where G_{i+1} is normal in G_i and $(G_i)/(G_{i+1})$ is abelian. Certainly an abelian torsion group is locally finite; hence, by successively applying Theorem D, we conclude that G_{n-1} is locally finite, G_{n-2} is locally finite, etc.

THEOREM F. The multiplicative group of all non-singular n by n triangular matrices over a field is solvable (and so is any subgroup).

We omit the proof.

For the proof of the next theorem, we need the following four lemmas, which we state without proof.

LEMMA 1. The cyclotomic polynomial of any order is irreducible over the rational numbers.

LEMMA 2. The degree $\phi(n)$ of the cyclotomic polynomial $\Phi_n(x)$ tends to ∞ with increasing n.

LEMMA 3. If f is a polynomial irreducible over a field F and F[x] is the field obtained from F by the adjunction of a transcendental element x, then f is irreducible over F[x].

LEMMA 4. Let K be a finite-dimensional extension of a field F. Then K can be represented (isomorphically) by matrices over F.

We shall also need the following.

LEMMA 5. Let G be a finitely generated torsion group of matrices over a field F. Then there exists a fixed integer N such that if α is a characteristic root of a matrix in G, then $\alpha^N = 1$.

(This is equivalent to saying that the elements of G are of bounded order; but we need only the statement about characteristic roots.)

Proof. Let $A_1, \ldots, A_r$ generate G. We may replace F by the field F_1, obtained by adjoining the elements of the matrices A_i to the prime field P of F. Then F_1 can be viewed as a finite-dimensional extension of a field Q, which is a purely transcendental extension of P. We use Lemma 4 (above) to write $A_1, \ldots, A_r$ as matrices over Q. Let us say the matrices (over Q) are n by n. Let α be a characteristic root of one of these matrices, and let h be the minimal positive integer for which $\alpha^h = 1$. By Lemma 3, the irreducible polynomial over Q which α satisfies is actually an irreducible polynomial over P. We wish to show that h is bounded for all α, and to do so, we argue in two cases.

Case I: If P is of characteristic zero, the irreducible polynomial for α is the cyclotomic polynomial Φ_h. Since the degree of Φ_h is less than or equal to n, the exponent h must be bounded (Lemma 2).

Case II: If P is of characteristic p, and α is of degree k over P, then the simple extension $P(\alpha)$ is the Galois field of p^k elements. Thus α to the power $p^k - 1$ is 1, so that $h \leq n$, i.e., h is bounded for all α.

Thus for some positive integer N and any characteristic root α, $\alpha^N = 1$.

THEOREM G. Any torsion group of matrices is locally finite.

Proof. Any torsion group of one by one matrices is locally finite. Assume that any torsion group of matrices of order less than n is locally finite. Let G^* be a torsion group of n by n matrices, and let G be a finitely generated subgroup of G^*.

Case I: If the underlying vector space is irreducible under G, then G is finite. For Lemma 5 implies that only a finite number of traces occur in the elements of G, so that G is finite by Theorem B.

Case II: If the space is reducible, each matrix in G can be written in the block form (see proof of Theorem C)

$$T = \begin{pmatrix} A & B \\ O & C \end{pmatrix}$$

The matrices A (and C) multiply independently. Thus, by the induction hypothesis, the set of matrices (A) is locally finite, and similarly for the set (C). We map G homomorphically onto the locally finite set of matrices

by

$$T \longrightarrow \begin{pmatrix} A & O \\ O & C \end{pmatrix}$$

The kernel of this homomorphism is the set of matrices

$$\begin{pmatrix} I_1 & B \\ O & I_2 \end{pmatrix}$$

where I_1 and I_2 are identity matrices. By Theorems E and F, this kernel is locally finite. Then Theorem D implies that G is locally finite. Consequently, G is finite.

It follows that G^* is locally finite.

3. Semi-Simple Rings

For any ring A, we shall use the symbol $R(A)$ for the radical of A. When no confusion is possible, we shall abbreviate $R(A)$ to R.

THEOREM 18. For any ring A, A/R is semi-simple.

Proof. We first establish the following lemma.

LEMMA. If $\overline{x}$ is a right quasi-regular element of A/R, and if x maps into $\overline{x}$ under the canonical homomorphism (modulo R), then x is right quasi-regular.

Proof. There is a $\overline{y}$ in A/R such that $\overline{x} \circ \overline{y} = 0$. Choose a y which maps into $\overline{y}$. Then $x \circ y$ is in R. But then x is R.Q.R., since there is a z such that

$$x \circ (y \circ z) = (x \circ y) \circ z = 0.$$

The statement of the theorem now follows readily; for, if R' is the radical of A/R, (by the Lemma) the inverse image of R' under the canonical homomorphism is a radical ideal in A. Such an ideal lies in R; hence $R' = (0)$.

Remarks. 1. If φ is a homomorphism of the ring A onto the ring B, then $\varphi(R(A))$ is contained in $R(B)$; however, $\varphi(R(A))$ is not in general all of $R(B)$.

2. Theorem 18 says that if one "divides out" the union of all radical ideals in a ring, the quotient ring has no (proper) radical ideals. To indicate that analogous propositions may fail, we remark that if one "divides out" the union of all nilpotent ideals in a ring, the quotient ring may have many nilpotent ideals. In fact, this quotient ring modulo its nilpotent ideal union may still have nilpotent ideals, and so on, ad infinitum.

For the proof of our next theorem, we need two lemmas.

LEMMA 19.1. In any ring A, if $-x^2$ is R.Q.R., then x is R.Q.R.

Proof. If $-x^2 \circ y = 0$, then

$$x \circ ((-x) \circ y) = x \circ (-x) \circ y = (-x^2) \circ y = 0.$$

LEMMA 19.2. Let A be a ring and x an element of A. Then xA is a radical ideal if and only if x lies in R(A).

Proof. If x is in R(A), certainly xA is a radical ideal. Suppose xA is a radical ideal, and let J be the right ideal generated by x. We shall prove J is a radical ideal. Now J is the collection of elements of the form $y = nx + xa$, where a is in A and n is an integer. One can easily check that for any such element y, $-y^2$ is in xA. By Lemma 19.1, y is R.Q.R.

THEOREM 19. Let I be a right ideal in a ring A. Then R(I) contains $I \cap R(A)$. If I is a two-ideal, then R(I) is equal to $I \cap R(A)$.

Proof. If x lies in $I \cap R(A)$, there is a y in A such that $x \circ y = 0$. Since $y = -x - xy$, y is in I. Thus $I \cap R(A)$ is a radical right ideal in I, and is contained in R(I). If I is in fact a two-ideal, then for any x in R(I) and any a in A, $-(xa)^2 = -x(axa)$ is in R(I) and is therefore R. Q. R. By Lemma 19.1, xa is R.Q.R. Thus xA is a radical right ideal in A, and by Lemma 19.2, x lies in R(A).

We now turn to the Chinese remainder theorem. The setting for this theorem is as follows. Let $I_1, \ldots, I_n$ be two-sided ideals in a ring A. There is then a natural homomorphism of A into the direct sum $A/I_1 \oplus A/I_2 \oplus \ldots \oplus A/I_n$. One may ask when this homomorphism is onto. If n = 2, the necessary and sufficient condition is that $I_1 + I_2 = A$, or, as we may say, I_1 and I_2 are _relatively prime_ ideals. The necessary and sufficient condition in general is that for $r = 1, 2, \ldots, n$, I_r be relatively prime to the intersection of the remaining I_k's. The Chinese remainder theorem states that, under suitable hypotheses, it is sufficient that I_k be relatively prime in pairs.

THEOREM (Chinese remainder). _Let_ $I_1, \ldots, I_n$ _be two-sided ideals in a ring_ A, _such that for_ $r \neq s$, $I_r + I_s = A$. _If each quotient ring_ A/I_r _is equal to its square (that is,_ $A^2 + I_r = A$), _then the natural homomorphism of_ A _into_ $A/I_1 \oplus \ldots \oplus A/I_n$ _is onto._

Proof. Since $I_1 + I_2 = A$, $I_1 + I_3 = A$,
$$A^2 = (I_1 + I_2)(I_1 + I_3) \subset I_1 + I_2 I_3 \subset I_1 + (I_2 \cap I_3),$$
the two inclusions being obvious. Since A/I_1 is equal to its square, $A = A^2 + I_1 \subset I_1 + (I_2 \cap I_3)$. Consequently, $A = I_1 + (I_2 \cap I_3)$. We repeat essentially the same argument using $A^2 = [I_1 + I_2 \cap I_3][I_1 + I_4]$ to conclude that $A = I_1 + (I_2 \cap I_3 \cap I_4)$. By induction, $I_1 + (I_2 \cap I_3 \cap \ldots \cap I_n) = A$. A similar argument shows that each I_r is relatively prime to the intersection of the I_k, $k \neq r$. This suffices to show that the mapping in question is onto.

Remark. The conditions $A^2 + I_r = A$ are automatically satisfied if each quotient ring A/I_r has a unit.

THEOREM 20. (Wedderburn-Artin). _Let_ A _be a semi-simple ring satisfying the descending chain condition on two-sided ideals. Suppose further that for every right primitive ideal_ P _in_ A, A/P _satisfies the descending chain condition on right ideals. Then_ A _is the direct sum of a finite number of simple rings, each of which is a full matrix ring over a division ring._

Proof. Let P be a right primitive ideal in A. By Theorem 17, A/P is a full matrix ring over a division ring; in particular A/P is simple and has a unit element. Thus the condition $A^2 + P = A$ holds. We now claim that there are only a finite number of right primitive ideals. For suppose we have an infinite number: $P_1, P_2, \ldots$. We claim that the chain $P_1, P_1 \cap P_2, \ldots, P_1 \cap \ldots \cap P_n, \ldots$ is properly descending. If, for instance,

$$P_1 \cap \ldots \cap P_n = P_1 \cap \ldots \cap P_n \cap P_{n+1},$$

then $P_1 \cap \ldots \cap P_n$ is contained in P_{n+1}. But by the Chinese remainder theorem, $P_{n+1} + (P_1 \cap \ldots \cap P_n) = A$. The conclusion is $P_{n+1} = A$, a contradiction. Our hypothesis that A satisfies the descending chain condition on two-sided ideals thus implies that the number of P_i's is finite. Their intersection is 0 by the semi-simplicity of A, and A is the full direct sum of the rings A/P_i by another application of the Chinese remainder theorem.

DEFINITION. Let A be a ring. The element a in A is called Regular (in the sense of von Neumann) if there is an x in A for which $axa = a$. If every element of A is Regular, then A is called a Regular ring.

Remark. Because of the many uses of the word "regular", we shall bend our grammar to keep our mathematics straight, and spell "von Neumann regular" with a capital "R".

Examples of Regular rings:

1. Any division ring is (obviously) Regular.

2. A direct sum of Regular rings is Regular.

3. The n by n matrix ring over a Regular ring is Regular (see Theorem 24). In particular, a semi-simple ring with D.C.C. on right (or left) ideals is Regular.

4. The ring of all linear transformations on a vector space V (not necessarily finite-dimensional) over a division ring is Regular.

Proof. Let A be a linear transformation on V, and let $\{v_\alpha\}$ be a basis for the range of A. Choose u_α such that $u_\alpha A = v_\alpha$. Then $\{u_\alpha\}$ together with the null space of A span V. As the v_α are linearly independent, there is a linear transformation X such that $v_\alpha X = u_\alpha$. Then $AXA = A$, as one can readily verify. We remark that the set of linear transformations on V with finite-dimensional range is a Regular subring of the full ring of transformations.

Remarks. 1. (von Neumann) In any Regular ring, the set of principal right ideals is a complemented, modular lattice. In his work on Continuous Geometry a big theorem is the converse: any complemented modular lattice satisfying certain mild conditions arises from a suitable Regular ring.

2. Any Regular Banach algebra is finite-dimensional.

3. A homomorphic image of a Regular ring is Regular.

THEOREM 21. Any Regular ring is semi-simple.

Proof. Let A be a Regular ring, and assume a is in $R(A)$. Suppose $axa = a$. Then $(ax)^2 = axa \cdot x = ax$, and ax is in $R(A)$. Thus $ax = 0$, since the radical of a ring never contains a non-zero idempotent. (If e is an idempotent in $R(A)$, then $-e$ is R.Q.R.; but, $-e + y - ey = 0$ implies $-e^2 + ey - ey = 0$, i.e., $e = 0$.)

LEMMA. (McCoy) If $axa - a$ is Regular, then a is Regular.

Proof.   If $(axa - a)y(axa - a) = axa - a$,   then
$a = a(x - y - xayax + xay + yax)a$.

THEOREM 22. Let A be a ring and I a two-ideal in A.
Then A is Regular if and only if both I and A/I are Regular.

Proof.   If A is Regular, we have already observed that
A/I is Regular. Let   a be in I. There is an x in A such
that axa = a. But then $a = a \cdot xax \cdot a$ and xax is in I. Thus I
is Regular. Now suppose that both I and A/I are Regular.
Let a be in A and let a map into a' under the canonical homo-
morphism of A onto A/I. There is an x' in A/I for which
$a'x'a' = a'$. Let x map into x' (modulo I). Then axa - a is in
I, and is therefore Regular. By McCoy's lemma, a is Regular.

DEFINITION. The two-ideal I in the ring A is called a
Regular ideal, if I is a Regular ring as a subring of A.

THEOREM 23.   Any ring A has a unique largest Regular
ideal M, and A/M has no non-zero Regular ideals.

Proof.   If I and J are Regular ideals in A, then I + J
is Regular. For, by the "first law of isomorphism" for rings,
$(I+J)/I$ is isomorphic to $J/(I \cap J)$. Now $J/(I \cap J)$ is Regular,
being a homomorphic image of the Regular ring J; since I is
Regular, I + J is Regular (Theorem 22). Evidently a finite sum
$I_1 + I_2 + \ldots + I_n$, of Regular ideals is also Regular.   Clearly
then the set M of all finite sums of elements, each lying in
some Regular ideal, is the unique largest Regular ideal in A.
Let J be a Regular ideal in A/M. The inverse image of J
under the homomorphism of A onto A/M is, by Theorem 22,
a Regular ideal in A, and therefore is contained in M. Consequently J = 0.

LEMMA 23.1. **If** I **is a Regular ideal in the ring** A **and** J **is a right ideal in** I, **then** J **is a right ideal in** A.

Proof. Let j be in J, a in A. There is an x in I with jxj = j. Then ja = jxja. Since xja is in I, ja is in J.

LEMMA 23.2. **Let** A **be a ring with a two-ideal** I. **If** I **has a unit, then** I **is a direct summand of** A.

Proof. Let e be the unit element for I. The key point is that e is central. To see this, we note that for any x in A, ex lies in I. It can therefore admit e as a right unit element, so we have ex = exe. In just the same way xe = exe. Hence ex = xe and e is central. If we let J be (1 - e)A = the set of all a - ea, we see readily that J is a two-sided ideal and that A is the ring direct sum of I and J.

Remark. If the ring A is a direct summand in every larger ring in which it is an ideal, then A has a unit.

Proof. Let B be the standard ring obtained by adjoining a unit to A. B is the set of ordered pairs (a, n), with a in A and n an integer. The operations in B are defined by $(a_1, n_1) + (a_2, n_2) = (a_1 + a_2, n_1 + n_2)$ and $(a_1, n_1) \cdot (a_2, n_2) = (a_1 a_2 + n_1 n_2 + n_2 a_1, n_1 n_2)$. The ring A is isomorphic to the ring of pairs $(a, 0)$, and A is thus an ideal in B. By hypothesis $B = A \oplus J$, for some ideal J contained in B. In particular, J contains an element (e, n) such that for some a in A, $(0, -1) = (a, 0) + (e, n)$. It must be that a = -e and n = -1. The significant point is that $(e, -1)$ is in J. Let b be in A. Then $(e, -1) \cdot (b, 0) = (eb - b, 0)$ is in both A and J, so eb = b. Similarly, be = b. Thus e is a unit for A.

THEOREM 23'. If the ring A has the descending chain condition on right ideals, it admits a direct sum decomposition $A = M \oplus N$, where M is the maximal Regular ideal of A, and N has no non-zero Regular ideals.

Proof. Lemma 23.1 guarantees that the maximal Regular ideal M of A (Theorem 23) satisfies the D.C.C on right ideals. Since any Regular ring is semi-simple, we see, from Theorem 20, that M has a unit. By Lemma 23.2, A admits a direct sum decomposition $A = M \oplus N$. Since N must be isomorphic to A/M, N has no non-zero Regular ideals (Theorem 23).

THEOREM 24. Any n by n matrix ring over a Regular ring is Regular.

Proof. We first consider the case $n = 2$. For the matrix $\begin{pmatrix} a & b \\ c & d \end{pmatrix}$, if $crc = c$, then

$$\begin{pmatrix} a & b \\ c & d \end{pmatrix} \begin{pmatrix} 0 & r \\ 0 & 0 \end{pmatrix} \begin{pmatrix} a & b \\ c & d \end{pmatrix} - \begin{pmatrix} a & b \\ c & d \end{pmatrix} = \begin{pmatrix} (arc-a) & (ard-b) \\ 0 & (crd-d) \end{pmatrix} .$$

Hence, by McCoy's lemma, it will suffice to consider matrices $\begin{pmatrix} a & b \\ 0 & d \end{pmatrix}$. Suppose that $axa = a$ and $dyd = d$. Then

$$\begin{pmatrix} a & b \\ 0 & d \end{pmatrix} \begin{pmatrix} x & 0 \\ 0 & y \end{pmatrix} \begin{pmatrix} a & b \\ 0 & d \end{pmatrix} - \begin{pmatrix} a & b \\ 0 & d \end{pmatrix} = \begin{pmatrix} 0 & axb-b \\ 0 & 0 \end{pmatrix} .$$

Again by McCoy's lemma, it suffices to consider matrices $\begin{pmatrix} 0 & b \\ 0 & 0 \end{pmatrix}$. But if $bzb = b$, then

$$\begin{pmatrix} 0 & b \\ 0 & 0 \end{pmatrix} \begin{pmatrix} 0 & 0 \\ z & 0 \end{pmatrix} \begin{pmatrix} 0 & b \\ 0 & 0 \end{pmatrix} = \begin{pmatrix} 0 & bzb \\ 0 & 0 \end{pmatrix} = \begin{pmatrix} 0 & b \\ 0 & 0 \end{pmatrix}$$

This concludes the argument for $n = 2$.

For $n = 4$, the conclusion follows immediately upon writing 4 by 4 matrices in the block form

$$\begin{pmatrix} A & B \\ C & D \end{pmatrix}$$

where A, B, C, D are 2 by 2 matrices. The 4 by 4 matrices are then simply 2 by 2 matrices over the ring of 2 by 2 matrices, and the above argument applies. By induction we easily obtain the result for $n = 2^k$.

For an arbitrary n, pick $2^k \geq n$. Then the ring of n by n matrices sits as the upper left n by n corner in the ring of 2^k by 2^k matrices. The desired result then follows from a remark which we leave as an exercise for the reader: if A is Regular and e is an idempotent in A, then eAe is Regular.

DEFINITION. Let G be a group and F a field. The group algebra of G over F is the set of all (formal) linear combinations $\Sigma \lambda_i g_i$ of the elements g_i of G, where the λ_i are elements of F, all but a finite number of which are zero. The algebraic operations are the obvious ones. (Use of group algebras has made it possible to exploit ring-theoretic techniques in the study of groups. We shall survey some of the known results and open questions, beginning with a complete result on semi-simplicity in case the group is finite.)

THEOREM 25. (Maschke) Let A be the group algebra of a finite group G over the field F. If F has characteristic zero, A is semi-simple. If F has characteristic p, then A is semi-simple if and only if p does not divide the order of G.

Proof. Of a number of different known proofs, we select one that uses a trace argument in the regular representation. The proof proceeds in two steps.

I. Suppose the characteristic p of F divides the order n of G. Let α be the sum of all the elements of G. Evidently $\alpha g = g\alpha = \alpha$ for any g in G. From this we deduce first that α lies in the center of the group algebra A. Furthermore, $\alpha^2 = n\alpha$ (n = order of G). Since p divides n, $\alpha^2 = 0$. Since α lies in the center of A, $(\alpha A)^2 = (0)$. Then αA, being a nilpotent ideal, lies in the radical of A (and is not (0)). Hence, A is not semi-simple.

II. Suppose F has characteristic zero or that it has characteristic p and p does not divide n. We need the concept of the <u>regular representation</u> of an algebra B. This representation is the homomorphic mapping of B onto the algebra of linear transformations over B (viewed as a vector space) determined by $x \rightarrow R_x$ where $a \cdot R_x = ax$. Note that when B has a unit this representation is faithful, i.e., an isomorphism.

We return to the group algebra A. Suppose $\Sigma \lambda_i g_i$ is a non-zero element of R(A). Multiplying by a suitable element of A, we obtain an element x in R(A) of the form $x = 1 + \mu_1 g_1 + \ldots + \mu_r g_r$. Under the regular representation of A, x maps into the linear transformation $R_x = I + \mu_1 R_{g_1} + \ldots + \mu_r R_{g_r}$. Choose as a basis of A the elements of G. Then each R_{g_i} is a permutation matrix corresponding to a permutation leaving no element fixed. It follows that R_{g_i} has zeros on the diagonal, so that the trace $Tr(R_{g_i}) = 0$. On the other hand x (being in the radical of a finite-dimensional algebra) is nilpotent; hence R_x is nilpotent; and hence $Tr(R_x) = 0$. But $Tr(R_x) = n \neq 0$ (p does not divide n). This is the desired contradiction.

We can deduce from Theorem 25 a result on the Regularity of certain infinite-dimensional group algebras.

THEOREM 26. <u>Let</u> G <u>be a locally finite group. Let</u> A <u>be the group algebra of</u> G <u>over a field</u> F. <u>If</u> F <u>has charac-teristic zero, or if</u> F <u>has characteristic</u> p <u>while no element of</u> G <u>is of order</u> p, <u>then</u> A <u>is Regular</u>.

<u>Proof</u>. Let x be in A. Then there exist elements $g_1, \ldots, g_n$ in G such that x lies in the subalgebra B generated by the g_i. As G is locally finite, $g_1, \ldots, g_n$ generate a finite subgroup G' of G. Now B is the group algebra of G' over F. By Theorem 25, B is semi-simple. But B is also Regular, as any finite-dimensional semi-simple algebra is Regular. Conse-quently, x is Regular.

This theorem stimulated a series of investigations concern-ing the validity of the converse. Partial results were obtained by M. Auslander, McLaughlin, and Villamayor. Then Villamayor (Pac. J. of Math. 9(1959), 941-951) proved the full converse: if the group algebra A of G over F is Regular then G is locally finite and, if F has characteristic p, G has no elements of order p.

We turn to the question of the semi-simplicity of a group algebra A, first over a field F of characteristic 0. A great deal of progress has been made by Amitsur and Herstein. The semi-simplicity of A is known if F is uncountable, or if F is transcendental over the rational numbers. To settle it in the re-maining cases (F algebraic over the rational numbers) it would suffice to handle the field of rational numbers. In any event, A has no nil ideals.

Similar results for characteristic p have been proved by Passman (Mich. Math. J. 9(1962), 375-384).

Analytic methods provide a quick proof of semi-simplicity over the field of complex numbers.

THEOREM H (Rickart, Segal)   <u>Any self-adjoint algebra</u> A <u>of bounded operators on a (complex) pre-Hilbert space is semi-simple</u>.

<u>Proof.</u>   We may as well assume the pre-Hilbert space complete (simply complete it!).

Let T be in $R(A)$, the radical of A. Then T^*T is in $R(A)$, and consequently, $-\lambda T^*T$ is quasi-regular for any complex number λ. If I is the identity operator, then $I - \lambda T^*T$ has an inverse in the full operator algebra; in other words, the spectrum of T^*T contains at most the number zero. It is known that this implies $T^*T = 0$. Therefore $T = 0$.

We can now demonstrate that the complex group algebra of any group G is semi-simple. For any $\alpha = \Sigma \lambda_i g_i$, g_i in G, λ_i complex, define

$$\|\alpha\|^2 = \Sigma |\lambda_i|^2 \ .$$

Then $\|\cdots\|$ is a norm, supporting an inner product, which makes the group algebra A into a pre-Hilbert space. To each element α in A make correspond the operator T_α on A for which $\beta T_\alpha = \beta \alpha$. The algebra of these operators T_α is self-adjoint : $T^*_\alpha = T_\gamma$, where $\gamma = \Sigma \overline{\lambda}_i g_i^{-1}$. Since A is evidently isomorphic to this operator algebra, we apply Theorem H and conclude that A is semi-simple.

At the other extreme, we study groups for which the group algebra is almost all radical. We introduce the <u>augmentation ideal</u> N: the set of all $\Sigma \lambda_i g_i$ with $\Sigma \lambda_i = 0$.

THEOREM 27. <u>If</u> G <u>is a finite</u> p-<u>group and</u> F <u>has characteristic</u> p, <u>then</u> N <u>is nilpotent.</u>

<u>Proof</u>. Of the several possible styles of proof, we choose to make an application of Kolchin's theorem (Theorem C in §2).

Fix a basis for A (as a vector space), and consider the regular representation $x \to R_x$. For each g in G there is a k such that $R_g^{p^k} = 1$. The characteristic roots of R_g are thus p^k-th roots of unity. Since F has characteristic p, these roots must in fact be equal to one. We now apply Kolchin's theorem and conclude that the matrices R_g can be simultaneously triangulated. Thus, the entire group algebra is triangulated in its regular representation. Consequently, N is the set of matrices with 0's down the main diagonal. Therefore N is nilpotent.

COROLLARY. <u>If</u> G <u>is a locally finite</u> p-<u>group and</u> F <u>has characteristic</u> p, <u>then</u> N <u>is a nil ideal</u>.

<u>Proof</u>. Let $x = \Sigma \lambda_i g_i$ be in N. We wish to show that x is nilpotent. Consider the finite subgroup of G generated by the g_i for which $\lambda_i \neq 0$. Applying Theorem 27, we see that x is nilpotent.

Actually, more is true: N is locally nilpotent (i.e., every finitely generated subalgebra of N is nilpotent). Losey (Mich. Math. J. 7(1960), 237-240) has proved the converse: if N is locally nilpotent then G is a locally finite p-group and F has characteristic p.

We are now going to lead up to a theorem on complete reducibility of representations which will make use of Theorem 26
and of the local finiteness of torsion matrix groups. First we
need some basic information on complete reducibility.

DEFINITION. The right A-module M is <u>completely re-
ducible</u>, if it is a direct sum of irreducible right A-modules.

THEOREM 28. <u>Let</u> A <u>be a ring, and suppose that</u> A <u>is
the sum of its minimal right ideals. If</u> A <u>contains no total</u> $(\neq 0)$
<u>left annihilator, then any right</u> A-<u>module</u> M <u>such that</u> MA = M
<u>is completely reducible.</u>

<u>Proof</u>. We first argue that M is a sum of irreducible
submodules. If x is in M, and I is a minimal right ideal in
A, consider the homomorphism of I onto xI which carries i
in I into xi. Since I is a minimal right ideal, it has no submodules (when viewed as a right A-module). Thus the kernel of
the above homomorphism is either (0) or I. If the kernel is I,
xI = 0. If the kernel is (0), xI $\cong$ I, an irreducible submodule
of M. (I is not trivial, that is, IA $\neq$ (0), since A has no total
left annihilators other than zero). The sum of all the modules
xI, as x runs through M, and I through the minimal right
ideal of A, is certainly MA = M. As the submodules xI = 0
may be disregarded, M is a sum of irreducible modules.

It remains to cut this sum down to a direct sum. This
calls for a straightforward transfinite induction (or use of Zorn's
lemma) quite analogous to picking a basis of a vector space, and
we leave it to the reader.

We need a lemma, whose proof we also leave to the reader.

LEMMA. If A is a semi-simple ring with the descending chain condition on right ideals, then A is the sum of its minimal right ideals.

DEFINITION. Let A be a finite-dimensional algebra with unit over a field F. Let V be a finite-dimensional vector space over F. By a representation of A on V we mean an algebraic homomorphism of A into the algebra of linear transformations on V, carrying the unit of A into the identity transformation.

Remark. A representation of A on V amounts to making V into a right A-module. Any submodule S of V is then automatically a subspace. If s is in S and λ in F, then $\lambda \cdot 1$ is in A, so that $\lambda s = s(\lambda \cdot 1)$, which lies in S.

THEOREM 29. Let G be a torsion group. Let V be a finite-dimensional vector space over a field F. If F has characteristic p, assume also that G has no elements of order p. Then any representation of G on V is completely reducible.

Proof. (By a representation of G on V we mean a homomorphism of G onto a multiplicative group of linear transformations over V, the identity of G mapping into the identity transformation. To say the representation is completely reducible is to say that V is a direct sum of subspaces which are invariant under the transformation group and are irreducible.) We first note that we may assume the representation to be faithful, i.e., an isomorphism; for, if this is not the case, we consider the isomorphism on G modulo the kernel of the representation. As G is a torsion group isomorphic to a matrix group, Burnside's theorem tells us that G is locally finite. Let A be the group algebra of A over F. By Theorem 26, A is Regular. The representation of G induces a representation of A. Let K be the kernel of this induced homomorphism (representation). Then A/K is Regular. But A/K is finite-dimensional. Therefore, A/K is semi-simple (Regularity implies semi-simplicity) and satisfies the descending chain condition on right ideals. It follows from the lemma and Theorem 28 that V is completely reducible.

There are numerous unsolved problems concerning group algebras. We shall conclude this section by mentioning three of them.

1. If G is torsion-free is its group algebra free of zero-divisors? Roughly speaking, the best that is known is that the answer is affirmative if G can be (linearly) ordered. This applies in particular to torsion-free abelian groups, for they can be ordered.

2. Let F be a field of characteristic 0, G an arbitrary group, and A the group algebra of G over F. Fact: if $xy = 1$ for x and y in A, then $yx = 1$. Moreover, the same is true for the algebra of n by n matrices over A. For an application of this result see Cockcroft and Swan (Proc. Lon. Math. Soc. 11 (1961), 194-202).

The only known proof is analytic. In brief: A can be embedded in a weakly closed algebra A_o (W^*-algebra, von Neumann algebra) of operators on the Hilbert space $L_2(G)$. It is known that A_o is "finite", a technical term in the theory of W^*-algebras which means precisely the property we are trying to prove. It is further known that finiteness of a W^*-algebra A_o implies finiteness of matrix algebras over A_o.

Open question: does this property (that one-sided inverses are two-sided) hold for group algebras in characteristic p ?

3. Again let F be a field of characteristic 0, G an arbitrary group, and A the group algebra of G over F. Let e be an idempotent in A, $e \neq 0$ or 1. Theorem (unpublished): the coefficient of the unit element in G is a totally real algebraic number with the property that it and all its conjugates lie strictly between 0 and 1. Again the proof is analytic. Query: is the coefficient in question actually rational?

4. The Wedderburn Principal Theorem

Let A be an algebra with radical R. The Wedderburn principal theorem asserts that under suitable hypotheses A is a vector space direct sum of R and a subalgebra S (necessarily S is isomorphic to A/R). This accomplishes a partial reduction of the study of algebras to the radical and semi-simple cases.

In this section we shall introduce an ad hoc hypothesis called "SBI" (hopefully some day there will be a more graceful name). We concentrate on the implications of this hypothesis for the lifting of idempotents.

The SBI idea arose in a conversation with Jacobson in the late 1940's. The intention was to unify a number of different contexts in which the lifting of idempotents is possible.

DEFINITION. Let A be a ring with radical R. We say that A is an SBI ring if for any y in R there exists an x in R such that

(1) $x^2 + x = y$,

(2) For all z in R, $yz = zy$ implies $xz = zx$ (i.e., x commutes with anything that commutes with y).

THEOREM 30. If R is nil, A is SBI.

Proof. We think of the formal solution of the quadratic equation $x^2 + x - y = 0$:

$$x = \frac{-1 + \sqrt{1 + 4y}}{2} \ .$$

If we expand $\sqrt{1 + 4y}$ by the binomial theorem and then simplify we find x to be a power series in y with integral coefficients. It starts

(*) $\qquad x = y - y^2 + 2y^3 - 5y^4 \ldots \ .$

Now if y is nilpotent, the series breaks off after a finite number of terms. The resulting element x certainly commutes with anything that commutes with y. We leave it to the reader to convince himself that we have solved the equation $x^2 + x = y$ (it is possible to formulate and prove rigorously a principle of the "preservation of identities").

Remarks. 1. Any Banach algebra is SBI. If y is in the radical, the series (*) converges.

2. Any compact topological ring is SBI, the above series again converging.

3. The following is an example of a ring which is not SBI. Let A be the ring of rational numbers with odd denominator. This is a principal ideal ring in which 2 is the only prime element. Accordingly, the radical is the single maximal ideal (2). For $y = -2$, no x (not even a real x) exists such that $x^2 + x = y$.

THEOREM 31. Let A be an SBI ring with radical R. Let u, v be orthogonal ($uv = vu = 0$) idempotents in A/R. If there exists an idempotent e in A which maps into u under the canonical homomorphism (of A onto A/R), then there exists an idempotent f in A, orthogonal to e, mapping into v.

Proof. In what follows, we shall make symbolic use of a unit. The reader may, if he so desires, simply think of adjoining a unit to A (if it does not already have one). Let b be an element of A mapping on v. Let $a = (1 - e)b(1 - e)$. Then a maps on v and $ea = ae = 0$. The element $z = a^2 - a$ lies in R, and $ez = ze = 0$. Consider $(2a - 1)^2 = 4a^2 - 4a + 1 = 4z + 1$. As z is in R, $(2a - 1)^{-2}$ exists (and commutes with z). Since A is SBI, there is a w in R such that $w^2 + w + z(2a - 1)^{-2} = 0$. This element commutes with e and a. Let $x = (1 - e)w$. Then

$x^2 + x + z(2a - 1)^{-2} = (1 - e)[w^2 + w + z(2a - 1)^{-2}] = 0$, and $ex = xe = 0$.

Let $r = x(2a - 1)$ and $f = a + r$. Since $ra = ar$,

$$f^2 = a^2 + 2ar + r^2 = a^2 + 2ax(2a - 1) + x^2(2a - 1)^2 .$$

Recalling that $x^2(2a - 1)^2 + x(2a - 1)^2 + z = 0$, we see that

$$f^2 = (a^2 - z) + 2ax(2a - 1) - x(2a - 1)^2 = a + x(2a - 1) = f .$$

Now $er = re = 0$, so that $ef = fe = 0$. Since r is in R, f maps onto v in A/R.

Remarks. 1. In connection with Theorem 31, we shall say that v has been lifted to an idempotent f in A.

2. If A is commutative, lifts of idempotents are unique. If e_1 and e_2 are both lifts of the same idempotent, $e_2 - e_2 e_1$ is an idempotent in the radical and is therefore zero. But then

$$(e_1 - e_2)^2 = e_1^2 - 2e_1 e_2 + e_2^2 = e_1 - e_1 e_2 + e_2 - e_1 e_2 = e_1 - e_2 ,$$

so $e_1 - e_2 = 0$.

3. Furthermore, when A is commutative, lifts of orthogonal idempotents are automatically orthogonal. If e_1 maps on u_1 and e_2 on u_2 where $u_1 u_2 = 0$ (u_i, e_i idempotent), then $e_1 e_2$ is an idempotent in the radical, and is thus zero.

COROLLARY. If A is an SBI ring and $u_1, u_2, \ldots$ are countably many orthogonal idempotents in A/R, then they can be lifted to orthogonal idempotents in A.

Proof. Taking $e = 0$ in Theorem 31, we can lift u_1 to an idempotent e_1 in A. With $e = e_1$, u_2 can be lifted to e_2, orthogonal to e_1. With $e = e_1 + e_2$, u_3 can be lifted to e_3, orthogonal to $e_1 + e_2$ (and consequently orthogonal to e_1 and e_2), etc.

Remark. The Corollary is in general false for an uncountable collection of orthogonal idempotents. For compact topological rings it is, however, true.

DEFINITION. Two idempotents e, f in a ring A are said to be <u>related</u> if there exist elements x in eAf and y in fAe such that $xy = e$ and $yx = f$. We note that e and f are related if and only if eA and fA are isomorphic right A-modules.

THEOREM 32. <u>Let</u> e <u>and</u> f <u>be idempotents in a ring</u> A, <u>mapping respectively into the elements</u> u <u>and</u> v <u>of</u> A/R <u>(where</u> R <u>is the radical of</u> A). <u>Then if</u> u <u>and</u> v <u>are related</u>, e <u>and</u> f <u>are related</u>.

<u>Proof.</u> By hypothesis, there exist elements $\overline{x}$ in $u(A/R)v$ and $\overline{y}$ in $v(A/R)u$ such that $\overline{x}\,\overline{y} = u$ and $\overline{y}\,\overline{x} = v$. Take any x_0 and y_0 mapping on $\overline{x}$ and $\overline{y}$, respectively. (We may assume x_0 in eAf and y_0 in fAe; if not, we would consider $ex_0 f$ and $fy_0 e$.) Now $x_0 y_0$ maps into u, so $x_0 y_0 - e$ is in R. There exists a t such that $x_0 y_0 - e + t + (x_0 y_0 - e)t = 0$. Multiplying this equation by e on the left (and recalling that $ex_0 = x_0$) we have $x_0 y_0 - e + x_0 y_0 t = 0$. Thus $x_0 y_0 + x_0 y_0 t = x_0 y_0 (e+t) = e$. Now let $x = x_0$ and $y = y_0 (e+t)e$. Certainly x lies in eAf and y in fAe. Also $xy = e$. To see that $yx = f$, we proceed as follows: yx is idempotent and is in fAf. Direct computation shows that yx maps into v in A/R. Consequently $f - yx$ is in R. But $f - yx$ is idempotent, and must therefore be zero.

The version we present of the Wedderburn principal theorem assumes that A/R is finite-dimensional and is a direct sum of total matrix algebras over the base field F, that is, the division algebras that might occur are assumed to collapse to F.

THEOREM 33. <u>Let</u> A <u>be an</u> SBI <u>algebra over a field</u> F, <u>such that</u> $A/R(A)$ <u>is finite-dimensional. Assume that each total matrix ring summand of</u> A/R <u>has</u> F <u>as its associated division ring. Then there exists a subalgebra</u> S <u>of</u> A <u>such that</u> A <u>is the vector space direct sum of</u> S <u>and</u> R.

<u>Proof</u>. Let $A/R = A_1 \oplus A_2 \oplus \ldots \oplus A_r$, where each A_i is a full matrix algebra over F. Let $\mu_{ij}^{(k)}$, $i, j = 1, \ldots, n_k$ be the matrix units of A_k. The elements $\mu_{ii}^{(k)}$ ($i = 1, \ldots, n_k$; $k = 1, \ldots, r$) are orthogonal idempotents in A/R which can (by Theorem 31) be lifted to orthogonal idempotents $e_{ii}^{(k)}$ in A. The elements $\mu_{11}^{(k)}$ and $\mu_{ii}^{(k)}$ are related, using $\overline{x} = \mu_{1i}^{(k)}$ and $\overline{y} = \mu_{i1}^{(k)}$ (the notation is that of Theorem 32). Therefore, by Theorem 32, $e_{11}^{(k)}$ and $e_{ii}^{(k)}$ are related for some $x = e_{1i}^{(k)}$ and $y = e_{i1}^{(k)}$. Now we define $e_{ij}^{(k)} = e_{i1}^{(k)} e_{1j}^{(k)}$. We must verify that this definition is consistent with the one above when $i = j$. But this follows immediately from the definition of $e_{i1}^{(k)}$ and $e_{1i}^{(k)}$. Now we claim that $e_{ij}^{(k)} e_{st}^{(k)} = \delta_{js} \cdot e_{it}^{(k)}$. By definition $e_{ij}^{(k)} e_{st}^{(k)} = e_{i1}^{(k)} e_{1j}^{(k)} e_{s1}^{(k)} e_{1t}^{(k)}$. Now $e_{1j}^{(k)}$ lies in $e_{11}^{(k)} A e_{jj}^{(k)}$ and $e_{s1}^{(k)}$ lies in $e_{ss}^{(k)} A e_{11}^{(k)}$. Thus, if $j \neq s$, $e_{1j}^{(k)} \cdot e_{s1}^{(k)} = 0$, and accordingly $e_{ij}^{(k)} e_{s1}^{(k)} = 0$. If $j = s$, then (by definition) $e_{1j}^{(k)} e_{s1}^{(k)} = e_{11}^{(k)}$. Recalling that $e_{i1}^{(k)} e_{11}^{(k)} = e_{i1}^{(k)}$ we obtain for $j = s$

$$e_{ij}^{(k)} e_{st}^{(k)} = e_{i1}^{(k)} e_{11}^{(k)} e_{1t}^{(k)} = e_{i1}^{(k)} e_{1t}^{(k)} = e_{it}^{(k)}$$

Now we know that $e_{ij}^{(k)} e_{st}^{(k)} = \delta_{js} e_{it}^{(k)}$; or, the $e_{ij}^{(k)}$ behave like matrix units.

Let S be the subspace of A spanned by all the $e_{ij}^{(k)}$, all i, j, k. Clearly S is a subalgebra. The image of S under the canonical homomorphism of A onto A/R is A/R, since the element

$$\sum_{i,j=1}^{n_1} \lambda_{ij}^{(1)} \mu_{ij}^{(1)} , \ldots, \sum_{i,j=1}^{n_r} \lambda_{ij}^{(r)} \mu_{ij}^{(r)}$$

of A/R is the image of the element $\sum_{i,j,k} \lambda_{ij}^{(k)} e_{ij}^{(k)}$ in S.

Furthermore $\sum_{i,j,k} \lambda_{ij}^{(k)} e_{ij}^{(k)}$ will have image zero only if

each $\lambda_{ij}^{(k)}$ is zero. That is $S \cap R = (0)$. It follows that A is

the vector space direct sum of S and R.

This is as far as we carry the theory. But we note that in classical accounts, or in contemporary versions using cohomology, the assumptions are stronger on R and weaker on A/R: it is assumed that R is nilpotent and that A/R is finite-dimensional and separable (i.e., the center of each simple component of A/R is a field separable over F). Separability cannot be dropped, as shown by the following example. Let F be a field of characteristic 2, containing an element c with no square root in F. Let A be the extension algebra $F[x]/(x^2+c)^2$ of F. Then A can be viewed as the algebra of polynomials in x of degree less than 4, multiplied with the restriction $x^4 = c^2$. The radical of A is the nilpotent ideal $R(A) = (x^2 + c)$. Then A/R is the field $F(\sqrt{c}) = F[y]/(y^2 + c)$. There is no subalgebra S of A such that A is the vector space direct sum of S and R, since this would imply S isomorphic to A/R, whereas S cannot contain a square root of c (since A does not).

As a final topic in this section we carry out the classification of two-dimensional algebras. There are eight possibilities (but note that the third one comprises all quadratic fields over F; as extreme cases there may be none at all or infinitely many).

Use of the Wedderburn principal theorem cuts down the work a trifle.

First we remark that the only one-dimensional algebras (associative or not) over a field F are F itself and a one-dimensional vector space with trivial multiplication. If A is a one-dimensional algebra generated by y, then $y^2 = \lambda y$ for some λ in F. If $\lambda = 0$, the multiplication in A is trivial; whereas if $\lambda \neq 0$, then the element $x = y/\lambda$ also generates A. Furthermore, $x^2 = x$, so that A is isomorphic to F.

Let A be a two-dimensional (associative) algebra over a field F. First, suppose that the radical R is all of A. Then since R is nilpotent, A^2 is either 0 (i.e., A is the trivial algebra), or A^2 is one-dimensional. Every element of A is nilpotent of index ≤ 3. In the second case we shall find an element which has index precisely 3. Select elements x, y in A with $xy \neq 0$. If either x^2 or $y^2 \neq 0$ we are done. But the assumption that both are 0 leads to a contradiction. For we can write $xy = \alpha x + \beta y$. Multiplying this equation on the left by x, we obtain $\beta xy = 0$, or $\beta = 0$. Multiplying on the right by y, we obtain $\alpha = 0$, which is absurd. Thus, one of the elements x, y has a non-zero square, call this element z. Now the nature of A is clear. The elements z and z^2 are linearly independent, since $z = \lambda z^2$ implies $z^2 = \lambda z^3 = 0$. Thus A consists of elements $\alpha z + \beta z^2$, multiplied with $z^3 = 0$.

Suppose, secondly, that $R = (0)$. Then A is a direct sum of total matrix rings over division rings. There are two possibilities:

3. If there is but one summand, A will be a division algebra over F (with unit). It follows that A is in fact a field, that is, a quadratic extension field of F.

4. If there are two summands, A is the direct sum of two copies of F.

Now let us consider the third possibility for R; namely, R is one-dimensional. We must have A/R isomorphic to F (being a one-dimensional semi-simple algebra over F). By the Wedderburn principal theorem, there exists a subalgebra S of A whose vector space direct sum with R is A. As S is isomorphic to A/R, it is isomorphic to F. Let e be the unit in S. Let u span A. The nature of A is then reflected completely in the products eu and ue. We have four possibilities.

5. If eu = ue = 0, then A is the direct sum of F and a one-dimensional trivial algebra over F.

6. Suppose eu = αu, $\alpha \neq 0$, while ue = 0. Then eeu = αeu. Recalling that e^2 = e, we have αu = α^2u, or α = 1. The multiplication table of A is e^2 = e, u^2 = 0, ue = 0, eu = u. It is convenient to picture A as the algebra of 2 by 2 matrices over F, having zeros in the second row,

$$e = \begin{pmatrix} 1 & 0 \\ 0 & 0 \end{pmatrix} \quad \text{and} \quad u = \begin{pmatrix} 0 & 1 \\ 0 & 0 \end{pmatrix}.$$

7. Another possibility is the dual of 6, namely ue = u and eu = 0.

8. The final possibility is eu = ue = u. This algebra is often called the algebra of dual members.

5. Theorems of Hopkins and Levitzki

After Artin introduced rings with descending chain condition there was a peaceful decade of development. The subject was thought of as distinct from (and in a sense dual to) the earlier theory of rings with ascending chain condition. It came as quite a surprise when Charles Hopkins proved (under the modest assumption of either a left or a right unit element) that the D.C.C. implies the A.C.C. We prove here a slightly sharpened version which has the merit of unifying his two theorems (note that the hypothesis of Theorem 34 is satisfied if A has either a left or a right unit element).

Let us first give an example to show that some hypothesis is needed. The example is one where multiplication is trivial (if this leaves the reader unsatisfied he is invited to explore draping a more complicated example around it). So: what we need is an abelian group with the descending but not the ascending chain condition on subgroups. The standard example is $Z(p^\infty)$: the additive group of rational numbers with denominator a power of the prime p, reduced modulo the subgroup of integers.

We state without proof the following theorem.

THEOREM. If G is an abelian group with descending chain condition on subgroups, then G is a direct sum of a finite group F and a finite number of groups $Z(p^\infty)$.

We shall need some preliminary results before proving Hopkins' theorem.

1) Let M be a right A-module. A chain is a descending (finite) set of submodules $S_1, \ldots, S_n$ of M:
$M \supset S_1 \supset S_2 \supset \ldots \supset S_n = (0)$. The factors of the chain are the quotient modules $M/S_1, S_1/S_2, \ldots, S_{n-1}/S_n$. Two chains are

equivalent if their factors are isomorphic in some order. We remind the reader of the principal result along these lines.

THEOREM. (Jordan-Hölder-Schreier-Zassenhaus) <u>Any two</u> <u>chains (of the same module) have equivalent refinements.</u>

A <u>composition series</u> is a maximal chain, i.e., one which cannot be properly refined. In a composition series the factors are either irreducible or trivial with a prime number of elements. The following result is deducible immediately from the theorem of Jordan-Hölder-Schreier-Zassenhaus.

THEOREM. <u>The right</u> A-<u>module</u> M <u>possesses a composi-</u> <u>tion series if and only if</u> M <u>satisfies both the ascending and de-</u> <u>scending chain conditions on submodules.</u>

2) <u>Lemma:</u> Let A <u>be a ring with unit</u> e, <u>and let</u> M <u>be a</u> <u>right</u> A-<u>module. Then</u> $M = S \oplus T$, <u>where</u> S, T <u>are submodules</u> <u>of</u> M, S <u>is unitary and</u> T <u>is trivial.</u>

<u>Proof.</u> Let $S = Me$, $T = M(1 - e)$, that is, T is the set of elements of M which are annihilated by e. Certainly S, T are submodules and T is trivial. We need only verify that $S \cap T = (0)$. But this is immediate, for if x is in S, $xe = x$, so that if x is also in T, $x = xe = 0$.

3) <u>Lemma:</u> Let A <u>be a ring with</u> D.C.C. <u>on the ideals of</u> <u>the form</u> $p^i A$, <u>where</u> p <u>is a prime. Let</u> x <u>be an element of</u> <u>order</u> p^r ($p^r x = 0$) <u>and infinite p-height, i.e., for any</u> n <u>there</u> <u>is a</u> y_n <u>with</u> $x = p^n y_n$. <u>Then</u> x <u>is a total annihilator in</u> A.

<u>Proof.</u> There is a k such that $p^k A = p^{k+1} A = B$. Then $B = pB = p^2 B = \ldots$, and $xB = p^r xB = (0)$. Let y be any element of A. Then $p^k y$ is in B, and we may write $p^k y = p^k z$, where z is in B. Now $xz = 0$. As x has infinite p-height, $x = p^k w$, for some w. Thus $x(y - z) = p^k w(y - z) = 0$, since $p^k y = p^k z$. Con-

sequently, $xy = xz = 0$. Since y was arbitrary, $xA = 0$. Similarly $Ax = 0$.

THEOREM 34. (Hopkins) <u>Let</u> A <u>be a ring with descending chain condition on right ideals. Let</u> R <u>be the radical of</u> A, <u>and suppose that for each non-negative integer</u> k <u>the ring</u> A/R^k <u>has no (non-zero) total annihilators. Then</u> A <u>satisfies the ascending chain condition on right ideals.</u>

<u>Proof.</u> We shall show that A, as a right A-module, has a composition series. As R is nilpotent, we may form a chain $A \supset R \supset R^2 \supset \dots \supset R^n = (0)$. We observe that it will suffice to prove that each module $A/R, R/R^2, \dots, R^{n-1}/R^n$ has a composition series. Each R^k/R^{k+1} is a right A-module, annihilated by R. Therefore, R^k/R^{k+1} can be regarded as a right A/R-module. Now A/R has a unit. By the lemma of 2) above, $R^k/R^{k+1} = S \oplus T$, where S is a unitary A/R-module and T is a trivial A/R-module. By Theorem 28, S is a direct sum of a finite number of irreducible submodules. The trivial module T has the D.C.C. on subgroups, and as we noted earlier, is therefore a direct sum of a finite group F and a finite number of groups $Z(p^\infty)$. If x is in $Z(p^\infty)$, $p^r x = 0$ for some r, and x has infinite p-height. Now x is in A/R^{k+1}. By the lemma 3) above, x is a total annihilator of A/R^{k+1}. By hypothesis, $x = 0$. Thus, $T = F$, a finite group. As R^k/R^{k+1} is a direct sum of a finite number of irreducible submodules and a finite group (trivial module) it certainly has a composition series.

<u>Remark.</u> By additional arguments one can refine this result; it suffices to assume that the ring itself has no non-zero total annihilators. See Fuchs, <u>Abelian Groups</u>, pp. 283-6.

We abruptly change subject matter to a pretty theorem due to Levitzki.

THEOREM 35. <u>Let</u> A <u>be the ring of all linear transforma-tions on an n-dimensional vector space</u> V <u>over a division ring. Let</u> S <u>be a multiplicative semi-group in</u> A, <u>consisting of nil-potent elements. Then the elements of</u> S <u>can be simultaneously put in strict triangular form, i.e., zeros on and below the main diagonal.</u>

<u>Proof.</u> We first remark that a particular conclusion from the theorem will be that the product of any n elements of S is zero. The proof will proceed by induction. For $n = 1$, it is trivial. Assume the result true for $n-1$. Note that if V is re-ducible under S, the elements of S may be represented in the block form $\begin{pmatrix} B & C \\ 0 & D \end{pmatrix}$, by a suitable choice of basis. The matrices B (the matrices D) constitute a semi-group of nilpotent linear transformations on a space of lower dimension, and may therefore be put simultaneously in strict triangular form, by vir-tue of the induction hypothesis. Thus S is strictly triangulable.

Let $T_1, \ldots, T_r$ be elements of S, and S' the semi-group generated by these elements. If V is irreducible under the semi-group S', then the vector space sum $VT_1 + VT_2 + \ldots + VT_r$ must either be (0) or V. If this sum is V, we may inductively select elements $T_{i_1}, T_{i_2}, \ldots$ from among the T_i such that $T_{i_k} \cdots T_{i_2} T_{i_1} \neq 0$, for each k. This is done as follows: if $T_{i_k} \cdots T_{i_2} T_{i_1} \neq 0$ then for some $i = 1, \ldots, r$, $T_i \cdot T_{i_k} \cdots T_{i_1} \neq 0$, or else $T_{i_k} \cdots T_{i_1}$ annihilates $V = VT_1 + \ldots + VT_r$. Take $T_{i_{k+1}} = T_i$. Now in the sequence $\{T_{i_k}\}$ some T_i, say T_1 for convenience, occurs at least $(n+1)$ times. We thus have a product $T_1 U_n T_1 U_{n-1} T_1 \cdots U_1 T_1 U_o \neq 0$, for some $U_o, \ldots, U_n$ in S'. Now the elements $U_i T_1$ for $i = 1, \ldots, n$ may be regarded as linear transformations of W, the range of T_1, into itself. As T_1 is nilpotent, the dimension of W is less than n. If we let S^* be the semi-group of transformations TT_1, T in S', we know from the induction hypothesis that S^* is strictly triangulable. As we have noted, it follows that $U_n T_1 \ldots U_2 T_1 \cdot U_1 T_1 = 0$, a contradiction. Thus V is irreducible under S' only if $S' = (0)$. But now it is clear that V cannot be irreducible under S, unless $S = (0)$. The case $S = (0)$ is trivial, and we have given the argument in case V is reducible under S.

COROLLARY. If a ring A satisfies the descending chain condition on right ideals, every multiplicative nil semi-group in A is nilpotent.

Proof. Let R be the radical of A. Then A/R is a direct sum of a finite number of total matrix rings over division rings. Let S' be the nil semi-group of A. The image S of S' in A/R is then a direct sum of a finite number of nil semi-groups of matrices over division rings. By Theorem 35, each of the latter semi-groups consists of matrices which can be simultaneously put in strict triangular form. Furthermore, if k is the maximum of the dimensions of the total matrix ring summands of A/R, the product of any k elements in S will be zero. For some n, $R^n = (0)$. It is easily seen that $(S')^{nk} = (0)$.

Remark. Theorem 35 should be compared with Kolchin's theorem (Theorem C in §2) which gives an analogous conclusion for a semi-group of unipotent matrices over a _field_. The proof of Kolchin's theorem could in fact be repeated to yield a different proof of Theorem 35, if the division ring were a field. It is an open question whether Kolchin's theorem holds over a division ring; at any rate a proof in the style of Theorem 35 does not appear to work. (Of course the division ring has to be infinite-dimensional over its center to have a new problem, for otherwise it can be represented by finite matrices over the center.)

We conclude this section with a theorem that simultaneously generalizes Kolchin's Theorem C and the field case of Theorem 35.

THEOREM H. <u>Let</u> S <u>be a multiplicative semi-group of matrices over a field</u> F. <u>Suppose each has the form</u> $\lambda I + N$ <u>for</u> λ <u>in</u> F <u>and</u> N <u>nilpotent. Then</u> S <u>can be put in simultaneous triangular form.</u>

Proof. A reduction to the algebraically closed case can be made just as in the proof of Theorem C; we therefore assume F algebraically closed. We may also assume the vector space V to be irreducible under S. As a final normalization it is harmless to assume that S contains all scalar matrices (enlarge S by taking the semi-group generated by S and all scalar matrices).

Let S_o be the subset of S consisting of its nilpotent matrices. S_o is closed under multiplication. Theorem 35 applies to show that S_o is triangular. In particular, there is a non-zero vector annihilated by S_o. Let W be the set of all x with $xS_o = 0$. For any $T \in S$, $T_o \in S_o$ we have that TT_o is singular and therefore nilpotent (i.e., in the sense of semi-group theory, S_o is an ideal in S). Hence $WTT_o = 0$. This proves that W is invariant under S. Since $W \neq 0$, we have $W = V$, $VS_o = 0$. We may therefore ignore S_o henceforth and assume S to consist of non-singular matrices.

Let S_1 be the subset of S consisting of matrices of determinant 1. (It is tempting to drop down further to the unipotent matrices; however we do not know this set to be closed under multiplication. Of course, when the proof finally concludes we will know this.) It is sufficient to triangulate S_1, for multiplication of any element of S by a suitable scalar throws it into S_1.

The characteristic roots of matrices in S_1 are n-th roots of 1, where n is the size of the matrices. Hence only finitely many traces occur in S_1, and S_1 is finite (Theorem B). This shows that S_1 is a group. We now make a case distinction, according to the characteristic of F.

Case I. Characteristic 0. If a matrix has finite multiplicative order and has the form scalar plus nilpotent, it must actually be a scalar. Thus S_1 just consists of scalars.

Case II. Characteristic p. A sufficiently high p^k-th power of each matrix in S_1 is a scalar, and in particular is central. This shows that S_1 is nilpotent and so is a direct product of groups of prime power order. The crucial point is that the elements of order a power of p form a subgroup, and they are exactly the unipotent matrices in S_1. They can be put in triangular form by Kolchin's theorem. Also any matrix in S_1 has the form scalar times unipotent. The proof of Theorem H is complete.

6. Primitive Rings with Minimal Ideals and Dual Vector Spaces

DEFINITION. Let D be a division ring, E a left vector vector space over D and F a right vector space over D. An inner product between E and F is a bilinear functional $(\cdot, \cdot)$ on $E \times F$ (values in D), i.e., if x_i is in E, y_i is in F, α, β in D, then

$$(x_1 + x_2, y_1) = (x_1, y_1) + (x_2, y_1)$$
$$(x_1, y_1 + y_2) = (x_1, y_1) + (x_1, y_2)$$
$$(\alpha x_1, y_1) = \alpha(x_1, y_1)$$
$$(x_1, y_1 \beta) = (x_1, y_1)\beta .$$

The spaces E and F are called dual if there exists a non-degenerate inner product between them, i.e., $(x, F) = 0$ implies $x = 0$; $(E, y) = 0$ implies $y = 0$.

DEFINITION. Let E be a left vector space over the division ring D. A linear functional on E is a linear mapping of E into D.

Remarks. 1. With obvious definitions, the set of all linear functionals on E becomes a right vector space over E. This space is called the full dual of E. It is actually dual to E, the inner product being defined by $(x, f) = f(x)$.

2. If F is any space dual to E, F is isomorphic to a subspace of the full dual of E. The element y in F corresponds to the functional f for which $f(x) = (x, y)$.

3. If E is finite-dimensional, any dual of E is the full dual of E.

4. If E is infinite-dimensional and F is the full dual of E, E is never the full dual of F.

We list examples of dual vector spaces, (E, F).

    1. F the full dual of E.

    2. E the full dual of F.

    3. If E, F have the same dimension, choose bases $(x_i), (y_i)$ for E, F and define an inner product by $(x_i, y_i) = \delta_{ij}$.

    4. Any Banach space E and its topological dual F.

    5. Form direct sums $E = E_1 \oplus E_2 \oplus \ldots$, $F = F_1 \oplus F_2 \oplus \ldots$, where (E_i, F_i) are dual, and define the inner product by $[e_1, e_2, \ldots; f_1, f_2, \ldots] = (e_1, f_1) + (e_2, f_2) + \ldots$.

DEFINITION. Let S be a subspace of E [subspace of F]. Then we denote by S' the set of elements y in F [x in E] such that $(S, y) = 0$ [$(x, S) = 0$]. We call $S'' = (S')'$ the <u>closure</u> of S, and say that S is closed if $S = S''$.

Remarks. 1. For any S, S' is closed. Hence, there is a one-one correspondence between the closed subspaces of E and those of F (a closed S corresponding to S'). This correspondence is an anti-isomorphism of the lattices of closed subspaces of E and F.

2. Every subspace of E is closed if and only if F is the full dual of E.

THEOREM. (Mackey) <u>If</u> E, F <u>are dual, then every finite-dimensional subspace of</u> E <u>is closed.</u>

Proof. If S is any subspace of E, S and F/S' are dual in a natural way. If x is in S, $\overline{y}$ in F/S' define $[x, \overline{y}] = (x, y)$ for some y in $\overline{y}$. This definition of $[x, \overline{y}]$ is independent of the y chosen, and yields a duality of S and F/S'.

Now let S be an n-dimensional subspace of E. Then F/S' being dual to S, is the full dual of S, and has thus dimension n. Similarly, S'' is dual to $F/S'' = F/S'$, and therefore has dimension n. It follows that $S = S''$.

Remarks. 1. The sum of two closed subspaces need not be closed; however, the sum of a closed subspace and a finite-dimensional subspace is always closed.

2. If E has countable dimension, and if the sum of any two closed subspaces of E is closed, then F is the full dual of E.

3. The sum of every two closed subspaces of E is closed if and only if the lattice of closed subspaces of E is modular.

THEOREM. (Mackey) If E and F are dual vector spaces of countable dimension, they admit dual bases.

Proof. Let $(u_1, u_2, \ldots)$ be a basis for E and $(v_1, v_2, \ldots)$ a basis for F. We wish to determine bases $(x_1, x_2, \ldots)$ for E and $(y_1, y_2, \ldots)$ for F such that $(x_i, y_j) = \delta_{ij}$. We proceed by inductive selection, distinguishing two cases: when n is even, and when n is odd. Suppose $x_1, \ldots, x_n$ and $y_1, \ldots, y_n$, linearly independent vectors in E and F, respectively, have been found such that $(x_i, y_j) = \delta_{ij}$. If n is even, proceed as follows. Let u_k be the first u_i linearly independent of $x_1, \ldots, x_n$. Let $x_{n+1} = u_k - (u_k, y_1)x_1 - \ldots - (u_k, y_n)x_n$. Then $(x_{n+1}, y_j) = 0$, $j = 1, \ldots, n$. Choose a vector w in F such that $(x_{n+1}, w) = 1$, and let $y_{n+1} = w - y_1(x_1, w) - \ldots - y_n(x_n, w)$. Then $(x_{n+1}, y_{n+1}) = 1$, and y_{n+1} is linearly independent of $y_1, \ldots, y_n$. This completes the induction step when n is even.

If n is odd, the process is essentially the same; however, in this case, one begins by selecting the first v_j which is linearly independent of $y_1, \ldots, y_n$. This alternating procedure guarantees that the elements $(x_i), (y_j)$ will span all of E and F respectively.

Remark. It follows from Mackey's theorem (in conjunction with Theorem 36) that there exists only one (up to isomorphism) simple algebraic algebra of countable dimension with a minimal one-sided ideal over an algebraically closed field.

DEFINITION. Let T be a linear transformation of E into E (E, F dual). The linear transformation T^* of F into F is called an <u>adjoint</u> of T if for every x in E and y in F:
$(xT, y) = (x, T^*y)$.

Remark. If an adjoint of T exists, it is necessarily unique.

THEOREM. <u>If</u> E <u>is a normed linear space, and</u> F <u>the</u> <u>topological dual of</u> E, <u>then</u> T <u>has an adjoint if and only if</u> T <u>is</u> <u>continuous.</u>

Proof. If T is continuous, the existence of an adjoint is easily demonstrated. Suppose T^* exists. To prove T continuous, it is enough to show that T is bounded on the unit sphere E_1 in E. For each y in F, $(E_1T, y) = (E_1, T^*y)$ is bounded. It follows from the Banach-Steinhaus uniform boundedness principle that E_1T is bounded.

Remark. If F is the full dual of E, every T has an adjoint; however, if E is the full dual of F, only special T's have adjoints.

We proceed now to the proof of Theorem 36, in which the close relation between dual vector spaces and primitive rings with a minimal ideal is demonstrated. For this purpose, we shall need some facts about minimal ideals and nilpotent ideals in a ring.

LEMMA 36.1. <u>If</u> I <u>is a nilpotent right ideal in a ring</u> A, <u>then</u> $I + AI$ (<u>the</u> 2-<u>sided ideal spanned by</u> I) <u>is a nilpotent</u> 2-<u>sided</u> <u>ideal.</u>

Proof. For a positive integer k, $(I+AI)^k = I^k + AI^k$. If $I^n = 0$, clearly $(I + AI)^n = 0$.

Remarks. 1. It is an elementary consequence of Lemma 36.1 that the statement "the ring A has no nilpotent ideals" is unambiguous; that is, the statements that A has no nilpotent left ideals, A has no nilpotent right ideals, and A has no nilpotent ideals (in all cases, "other than (0)") are equivalent.

2. It is an open question whether the corresponding result is valid for nil ideals.

LEMMA 36.2. *If* e *is an idempotent in a ring* A *and* eA *is a minimal right ideal, then* eAe *is a division ring.*

Proof. Note that e is a 2-sided unit for eAe. Suppose exe $\neq$ 0. Then exeA is a non-zero right ideal contained in eA. Thus, exeA = eA. For some y then exey = e, or exe·eye = e. Thus, eAe is a division ring.

LEMMA 36.3. *Let* A *be a ring with no (non-zero) nilpotent ideals. If* e *is an idempotent in* A *such that* eAe *is a division ring, then* eA *is a minimal right ideal.*

Proof. Let I be a non-zero ideal contained in eA. Let ex $\neq$ 0 be in I. Consider exAe. Suppose exAe = 0. Then exAexA = (0), i.e., exA is a nilpotent right ideal. By hypothesis exA = 0. Now the set of total left annihilators of A is a nilpotent ideal (and contains ex); hence, ex = 0, a contradiction. Thus, exAe $\neq$ (0). Then, for some a in A, there is a b such that exae·ebe = e. But exae·ebe is in I. Therefore I contains e, and consequently must be equal to eA.

LEMMA 36.4. *If* I *is a minimal right ideal in a ring* A, *then either* $I^2 = 0$, *or* I = eA, e *an idempotent.*

Proof. Assume $I^2 \neq (0)$, and choose an element a in I such that aI $\neq$ (0). Then aI = I. In particular, there is an element e in I for which ae = a. The right annihilator, in I, of the element a is a right ideal contained in I. As it is not I (it does not contain e), the annihilator is (0). Now $ae^2 - ae$ $= a(e^2 - e) = 0$; hence $e^2 = e$. Consider the right ideal $eA \subset I$. Certainly $eA \neq (0)$, as $e \cdot e = e \neq 0$ is in eA. It follows that $I = eA$.

Notation: If E, F are dual vector spaces, let $L = L(E, F)$ denote the ring of continuous linear transformations over E. Let $S = S(E, F)$ denote the subring of L, consisting of those transformations of finite-dimensional range.

THEOREM 36. Let E and F be dual vector spaces. Let A be any ring containing $S = S(E, F)$ and contained in $L = L(E, F)$. Then S is the unique minimal 2-ideal in A. Also, A is (left and right) primitive, and has a minimal left ideal. Conversely, any primitive ring with a minimal left ideal arises in this way from a pair of dual vector spaces.

Proof. Let $S \subset A \subset L$. First we note that if x_o is in E and y_o in F, the mapping: x into $(x, y_o)x_o$ is a linear transformation T of one-dimensional range. Also, T is continuous: $T^*y = y_o(x_o, y)$. Furthermore, any continuous T of one-dimensional range arises in this way. Let x_o be a fixed non-zero vector in E, and consider the set I of elements of L which map E into the one-dimensional space spanned by x_o. A typical element T in I is obtained by fixing a y in F and setting $xT = (x, y)x_o$. Certainly I is a left ideal in A. Moreover, I is a minimal left ideal in A; for, if a left ideal contains T_o : $xT_o = (x, y_o)x_o$, $y_o \neq 0$, it contains every T in I, since $T = T_1 T_o$, where $xT_1 = (x, y)x_1$, and x_1 is so chosen that $(x_1, y_o) = 1$. It follows that A is (left) primitive. In the second half of the proof, we shall see that A is also right primitive. It remains only to show that S is the unique minimal 2-ideal in A. Clearly any non-zero right ideal intersects S, so that there can be no minimal 2-ideal other than S. To prove that S is minimal, it will suffice to show that a non-zero 2-ideal in S must contain every transformation (continuous) of one-dimensional range (each element of S is a sum of such transformations). Let T be a non-zero element of A, say $uT = v \neq 0$. Let T_0 be an arbitrary element of S of one-dimensional range: $xT_0 = (x, y_0)x_0$. Define $xT_1 = (x, y_0)u$ and let T_2 be such that $vT_2 = x_0$. Then $T_0 = T_1 T T_2$. Thus S is minimal.

Now suppose that A is a primitive ring with minimal left ideal I. (We shall see shortly that for rings with minimal left ideal, left and right primitive are the same.) By Lemma 36.4, $I = Ae$, e an idempotent. (A primitive ring has no non-zero nilpotent ideals.) By Lemma 36.2, eAe is a division ring. By Lemma 36.3, eA is a minimal right ideal. Now eA is a left vector space over eAe and Ae is a right vector space over eAe. We define the following inner product between eA and Ae; $(ex, ye) = exye$. The linearity properties are obvious, and non-degeneracy is demonstrated as follows. If $(ex, Ae) = 0$, i.e., if $ex \cdot Ae = 0$, then $(exA)^2 = 0$, implying that $exA = 0$ and $ex = 0$. (The argument for elements ye is similar.)

With each element x in A we associate the linear transformation T_x on eA defined by $(ea)T_x = eax$. Now T_x is continuous, T_x^* being left multiplication by x on Ae. The mapping of x into T_x is clearly a homomorphism. It is also easily seen to be one-one. We now have A isomorphic to a subring of $L(eA, Ae)$. It remains only to show that A contains every element of $S(eA, Ae)$. For this, it suffices to show that A contains every element of S having one-dimensional range. But this is immediate; for if $(ea)T = (ea, ey)ex$, then T is simply right multiplication by yex.

Remarks. 1. A consequence of the above proof is that if a (left or right) primitive ring A contains either a left or right minimal ideal, it is both left and right primitive and contains both left and right minimal ideals. All faithful irreducible right A-modules are isomorphic.

2. An elementary consequence of Theorem 36 is that a simple ring A containing a minimal left ideal I (where I = Ae as above) is an S(E, F), namely S(eA, Ae). One may consider simple rings as being of two types: those with a minimal left (or right) ideal, and those without such an ideal. Those of the first type are completely described (above) , whereas knowledge of those of the second type is scanty. We have seen one example of a simple ring without a minimal left ideal (the ring of "differential polynomials"). Another example can be constructed as follows. Let L be the ring of linear transformations on a countable-dimensional vector space, and let I be the ideal of those of finite-dimensional range. Then A = L/I is simple without a minimal left ideal.

7. Simple Rings

(1) The enveloping ring and the centroid

Up to Theorem 44 the associative law will be irrelevant. Since there are important applications (notably to Lie algebras) we drop associativity at this point.

Let A be any ring. Associated with an element x in A we have the right and left multiplications

$$R(x): a \to ax$$

$$L(x): a \to xa .$$

The ring E generated by all L's and R's is called the enveloping ring of A; it is a subring of the ring of endomorphisms of the abelian group A. The general element of E is a sum of terms, each of which is a product of L's and R's. When E is associative, the general element of E takes on the simpler form

$$L(a) + R(b) + \Sigma \ L(c_i) R(d_i) .$$

We think of the elements of E as placed on the right of A and in this way A becomes a right E-module. Examining the relevant definitions we see that A is simple if and only if it is an irreducible E-module.

For later use we state at this point:

THEOREM 38. _Let_ a _be a non-zero element of a simple ring_ A. _Then for a suitable integer_ n $(n = 0, 1, 2, ...)$ _and elements_ $x_1, ..., x_n$, y _in_ A _we have_ $aR(x_1)...R(x_n)L(y) \neq 0$.

Proof. Suppose the contrary. Write I for the set of all elements of the form

$$ka + \Sigma \ aR(x_1)...R(x_j) ,$$

where k is an integer. Then I is invariant under right multiplication. Also $IL(y) = 0$ for any y. Hence I is a two-sided ideal and it is non-zero since it contains a. Thus $I = A$. But then $yA = 0$ for all y, $A^2 = 0$, a contradiction.

The <u>centroid</u> C of a ring A is the ring of all additive endomorphisms of A which commute with all L's and R's. Thus if S is in the centroid we have

$$(xy)S = (xS)y = x(yS)$$

for all x, y in A.

<u>Examples:</u>

1. If A is trivial, the centroid is the full ring of endomorphisms of A.

2. If A has a unit element, it can be seen that the centroid coincides with the ordinary center, i.e., the set of elements commuting and associating with everything.

3. If A is an algebra over a field F, the centroid contains F and is itself an algebra over F.

4. As a generalization of the Gelfand-Mazur theorem, one can prove that the centroid of a primitive Banach algebra is just the complex numbers.

THEOREM 39. <u>For any elements</u> x, y <u>in a ring</u> A <u>and any centroid elements</u> S, T <u>we have</u> $(xy)ST = (xy)TS$.

<u>Proof.</u>

$$(xy)ST = (x \cdot yS)T = xT \cdot yS$$
$$= (xT \cdot y)S = (xy)TS.$$

THEOREM 40. <u>If</u> $A^2 = A$, <u>the centroid of</u> A <u>is commutative. If</u> A <u>has no non-zero total annihilator, the centroid is commutative.</u>

<u>Proof.</u> The first part is immediate from Theorem 39. To prove the second part we note

$$x \cdot y(ST - TS) = (xy)(ST - TS) = 0$$
$$y(ST - TS) \cdot x = (yx)(ST - TS) = 0$$

Thus $y(ST - TS)$ is a total annihilator of A and is 0.

THEOREM 41. The centroid of a primitive associative ring is an integral domain.

Proof. The commutativity follows from Theorem 40. We omit the proof that there are no divisors of 0.

THEOREM 42. The centroid of a simple ring is a field.

Proof. The centroid is a division ring by Schur's lemma and is commutative by either half of Theorem 40.

It follows from Theorem 42 that any simple ring can be regarded as an algebra, for instance over its centroid. If the base field is exactly the centroid we call the algebra central simple.

To complete the identification of the concepts of simple ring and simple algebra, we should also note that if an algebra A is simple in the sense of having no algebra ideals, then it is also simple as a ring. For let J be a non-zero ring ideal in A. Let I be the subspace spanned by J. Then I is an algebra ideal so that $I = A$. The typical element of I is of the form $\Sigma \lambda_i a_i$, a_i in J. Then $(\Sigma \lambda_i a_i)x = \Sigma a_i(\lambda_i x)$ is in J. Hence J contains IA, which in turn is $A^2 = A$.

(More generally, if a module is irreducible when operators are allowed, it is also irreducible without operators.)

For associative rings it is convenient to introduce the reduced enveloping ring E', defined as the set of all sums $\Sigma L(x_i)R(y_i)$. With a unit element, this is the same as the enveloping ring.

THEOREM 43. <u>Let</u> A <u>be a simple associative ring,</u> E'
<u>its reduced enveloping ring. Then</u> A <u>is also irreducible as an</u>
E'<u>-module, and the commuting ring of endomorphisms is again</u>
<u>the centroid.</u>

We omit the proof.

(2) Tensor products

In the interst of speed we define the tensor product by means
of bases, although we freely acknowledge that an invariant definition is to be preferred.

Let A, B be algebras over a field F. Let $\{u_i\}, \{v_j\}$ be
bases of A, B. We define $A \otimes B$ to be an algebra with basis
$u_i v_j$ (or more cautiously we might write $u_i \otimes v_j$) with multiplication table

$$u_i v_j \cdot u_k v_\ell = u_i u_k \otimes v_j v_\ell \ ,$$

where of course the right hand side is to be expanded by the dis-
tributive law.

The general element of $A \otimes B$ is thus of the form
$\Sigma \, \alpha_{ij} u_i v_j$, $\alpha_{ij} \in F$. If we gather all terms involving v_j in the
single term a_j , we may rewrite it as $\Sigma \, a_j v_j$. In other words:
we use a basis of B with coefficients ranging over A (instead
of F). Since this description no longer utilizes a basis of A,
the tensor product is independent of the choice of basis.

Examples:

1. If K is an extension field of F, $K \otimes A$ is an algebra
over K, with the same basis and multiplication table as the origi-
nal algebra. One speaks of extending the base field from F to K.

2. Let Q be the quaternions as an algebra over the reals,
and K the complex numbers. Then $K \otimes Q$ is the two by two
total matrix algebra over K.

3. $Q \otimes Q$ is the 4 by 4 total matrix algebra over the reals.

4. $K \otimes K$ is the direct sum of two copies of the complex numbers.

5. If A is any algebra over F, and M_n is the n by n total matrix algebra over F, then $A \otimes M_n$ is the n by n total matrix algebra over A.

6. $M_r \otimes M_s$ is isomorphic to M_{rs}.

THEOREM 44. <u>Let</u> A <u>and</u> B <u>be algebras over</u> F. <u>Sup-pose that</u> A <u>is central simple and</u> B <u>is simple.</u> <u>Assume further any one of the following three hypotheses:</u>

(1) A <u>has a unit element,</u>

(2) B <u>has a unit element,</u>

(3) A <u>is associative and</u> B <u>has no non-zero total left or right annihilators.</u>

<u>Then</u> $A \otimes B$ <u>is simple.</u>

Before presenting the proof, let us give an example to show that some hypothesis is necessary. Let A be the two-dimensional algebra with basis u, v and table $u^2 = uv = 0$, $vu = v$, $v^2 = u$. You can check simplicity of A rapidly by noting:

$$R_v = \begin{pmatrix} 0 & 0 \\ 1 & 0 \end{pmatrix} \quad , \quad L_v = \begin{pmatrix} 0 & 1 \\ 1 & 0 \end{pmatrix} ,$$

and that R_v and L_v suffice to generate all two by two matrices. A is in fact central simple, but we can bypass this point by assuming the base field to be algebraically closed. Note that u is a left annihilator of A. Let B be the algebra anti-isomorphic to A. Then B contains a right annihilator, say u_1. In $A \otimes B$ the element uu_1 is thus a two-sided annihilator and gives rise to a one-dimensional ideal.

<u>Proof of Theorem 44</u>. Let I be a non-zero ideal in $A \otimes B$. We prove that $I = A \otimes B$, dividing the proof into three parts.

I. Suppose I contains Ab for some non-zero b in B. Right multiplying by ab_1 we see that I contains $Aa \cdot bb_1$. This being true for any a in A we get $AA \cdot bb_1 \subset I$. Since A is simple we have $A^2 = A$, $A \cdot bb_1 \subset I$. In this way we can build up the two-sided ideal generated by b, which is all of B since B is simple. Hence $I = A \otimes B$.

II. Suppose I contains a non-zero element ab. Central simplicity of A is irrelevant here and so we may treat hypotheses (1) and (2) together, supposing that B has a unit element. We can carry out left and right multiplications on a, holding b fixed. When this is followed by additions we get all of Ab in I. The argument then reverts to Case I. Under hypothesis (3) we pick elements b_1, b_2 in B such that $b_1 b \cdot b_2 \neq 0$. Then for any x, y in A we have $(xay)(b_1 b \cdot b_2) \in I$. Since $AaA = A$, by adding such terms we find $A(b_1 b \cdot b_2) \subset I$. We refer again to Case I.

III. In the general case we begin with a non-zero element $x = \Sigma a_i b_i$ in I. We may suppose that the a's are linearly independent over F, and the b's non-zero. By the density theorem there exists an element T in the enveloping ring of A such that $a_1 T \neq 0$, $a_i T = 0$ for $i > 1$; if A is associative we can pick T in the reduced enveloping ring (Theorem 43). We now distinguish the three hypotheses.

(2) If B has a unit element, the left and right multiplications that build up T can be carried out on A while leaving the B-component fixed. Applying these to x we get that $a_1 T \cdot b_1$ lies in I, and we refer to Case II.

(1) Suppose that A has a unit element. Let us consider formal products of L's and R's with no symbols yet attached to them. We speak of one such product as being a <u>refinement of</u>

of another if it is obtained from it by the insertion of more L's and R's. By repeated applications of Theorem 38 we can find an element U of the enveloping ring of B with the following properties: (1) U is just a single product of L's and R's. (2) $b_1 U \neq 0$, (3) U is a refinement simultaneously of each of the monomials comprising T. Now since A has a unit element we can stick into each of these monomials harmless left and right multiplications by 1. The result is to make T a sum of terms, each having the same formal product of L's and R's as U does. This makes it possible to apply TU to the element $x = \Sigma\, a_i b_i$. Since each $a_i T = 0$ for $i \geq 2$, we conclude that $a_1 T \cdot b_1 U$ is in I. This reverts the problem to Case II.

(3) We must finally treat the case where A is associative and B has no left or right annihilators. We can find c, d in B such that $cb_1 \cdot d \neq 0$. Since T is now in the reduced enveloping ring, it has the form $T = \Sigma\, L(u_j) R(u_j)$. Left multiply $x = \Sigma\, a_i b_i$ by $u_j c$, then right multiply by $v_j d$, then add over j. The result is that $a_1 T \cdot (cb_1 \cdot d)$ lies in I. This completes the proof of Theorem 44.

From now on all rings are again associative.

As a first application of Theorem 44, consider a division algebra D, finite-dimensional over its center F. We claim that [D:F] is a square. For let K be an algebraically closed field containing F. We form $K \otimes D$. Since K is simple and D is central simple, $K \otimes D$ is simple by Theorem 44. Moreover $K \otimes D$ is an algebra over K, with the same dimension as D over F. Thus $K \otimes D$ is a total matrix algebra, its dimension over K is a square, and [D:F] is a square.

THEOREM 45. (The internal tensor product theorem) <u>Let</u> C <u>be an algebra over</u> F. <u>Let</u> A, B <u>be subalgebras with</u> A <u>cen-</u>

tral simple and B simple. Suppose further that A and B commute elementwise. Then AB is either 0 or is isomorphic to $A \otimes B$.

Proof. There is a natural homomorphism from $A \otimes B$ onto AB. By Theorem 44, $A \otimes B$ is simple. Hence the kernel is either 0 or all of $A \otimes B$.

THEOREM 46. Let A be a central simple algebra over F, and denote the reciprocal algebra by A^*. Then $A \otimes A^*$ is a dense ring of linear transformations on a vector space (namely A) over F.

Proof. In the algebra E of all linear transformations on the vector space A we observe two subalgebras: A_r, A_ℓ the algebras of right and left multiplications, respectively, by elements of A. A_r is isomorphic to A, A_ℓ is isomorphic to A^*. A_r and A_ℓ commute elementwise by the associative law. Finally $A_r A_\ell$ is obviously non-zero. Hence (Theorem 45) $A_r A_\ell \cong A \otimes A^*$. Now $A_r A_\ell$ is exactly the reduced enveloping ring of A. By Theorem 43, $A_r A_\ell$ is a dense algebra of linear transformations on A, as a vector space over the centroid F.

In particular: if A is central simple finite-dimensional, then $A \otimes A^*$ is a total matrix algebra. This is the starting point for making a group (the Brauer group) out of the central simple finite-dimensional algebras.

(3) Maximal subfields

Any ring possesses maximal commutative subrings by Zorn's lemma. Note that they necessarily contain the center. In a division ring a maximal commutative subring is automatically a subfield.

THEOREM 47. <u>Let</u> D <u>be a central division algebra over</u> F, <u>and</u> K <u>a maximal subfield. Then</u> K$\otimes$D <u>is a dense ring of linear transformations on a vector space over</u> K (<u>namely</u> D <u>as a left vector space over</u> K).

<u>Proof</u>. We operate again in the algebra E of all linear transformations on D as a vector space over F. Let D_r denote the algebra of all right multiplications by D, K_ℓ the algebra of left multiplications by K. By Theorem 45, $K_\ell D_r \cong K\otimes D$. Let us look at D as a right $K_\ell D_r$-module. It is irreducible, being already irreducible under D_r. What is the commuting division ring? One readily computes that an endomorphism commuting with D_r is of the form L_x with x in D. For L_x to commute with K_ℓ, it must further be the case that x commutes with K. By the maximality of K, x lies in K. Thus the commuting division ring is exactly K_ℓ. The density theorem completes the proof.

If in particular D is finite-dimensional over F, then K$\otimes$D is a total matrix algebra over K and we see once again that [D:F] is a square. But we can get more precise information.

THEOREM 48. <u>Let</u> D <u>be a division algebra over</u> F, <u>and</u> A <u>a finite-dimensional algebra with unit element over</u> F. <u>Then</u> A$\otimes$D <u>satisfies the descending chain condition on right ideals</u>.

<u>Proof</u>. We first look at A$\otimes$D as a right vector space over D, with D acting in the natural way on the right. It is a finite-dimensional vector space: in fact if $u_1, \ldots, u_r$ is a basis of A over F then $u_1, \ldots, u_r$ is also a basis of A$\otimes$D over D. In particular A$\otimes$D satisfies the descending chain condition on D-submodules.

Now look at $A \otimes D$ as a right $(A \otimes D)$-module. The elements $1 \cdot d$ of $A \otimes D$ act on the right of $A \otimes D$ in just the way we had D acting in the preceding paragraph. Hence an $(A \otimes D)$-submodule (that is, a right ideal of $A \otimes D$) is a D-subspace. A fortiori, we have the descending chain condition on right ideals of $A \otimes D$.

THEOREM 49. Let D be a central division algebra over F, K a maximal subfield. If D is infinite-dimensional over F, so is K. If D is finite-dimensional over F, its dimension is a square n^2, and $[K:F] = n$.

Proof. Suppose that K is finite-dimensional over F, say $[K:F] = r$. By Theorem 48, $K \otimes D$ satisfies the descending chain condition. We now apply Theorem 47 and observe that D, as a left vector space over K, must be finite-dimensional (for a ring with descending chain condition acts as a dense ring on it). If $[D:K]$ is s, then $D \otimes K$ is an s by s total matrix algebra over K. Thus $[D:F] = s^2$. But on the other hand $[D:F] = [D:K][K:F]$ $= sr$, so that $r = s$. We have proved both statements of Theorem 49.

As an application of Theorem 49, let D be an algebraic division algebra over a real-closed field R. Then a maximal subfield must be R or the complexes. We rapidly conclude that D is the reals, complexes, or four-dimensional over R. We shall complete the determination of D a little later, with the aid of more theory.

(4) Polynomial identities

Let C be a commutative ring. Let $x_1, \ldots, x_n$ be indeterminates. A non-commutative monomial is a product of x's, order being carefully observed. A non-commutative polynomial over C is a linear combination of monomials with coefficients in C. We do not allow a constant term.

Let A be a ring admitting C as a ring of operators (in other words, C is part of the centroid). We say that A satisfies a polynomial identity (over C) if there exists a non-zero non-commutative polynomial over C which vanishes whenever elements of A are substituted.

Examples:

1. Any nil ring of bounded index.

2. Any commutative ring.

3. Any finite-dimensional algebra. If the dimension is k-1, the identity

$$S(x_1, \ldots, x_k) = \Sigma \pm x_{i_1} x_{i_2} \ldots x_{i_k}$$

is satisfied, the sum being over all permutations with the sign according to the parity of the permutation.

4. Any algebraic algebra of bounded degree. If the bound on the degree is n, the identity

$$S(xy, x^2 y, \ldots, x^n y) = 0$$

is satisfied.

THEOREM 50. Let A be a primitive ring satisfying a polynomial identity of degree k, with coefficients in the centroid. Then A is a simple algebra, finite-dimensional over its center. If the dimension over the center is n^2, then $2n \leq k$.

First a preliminary lemma.

LEMMA. Let I be a maximal right ideal in a ring A. Assume that I does not contain A^2. Then I is invariant under the centroid of A.

Proof. Let λ be a centroid element, placed on the left. Evidently λI is a right ideal in A. If λI is not contained in I, then $I + \lambda I = A$. Then

$$A^2 = IA + \lambda IA = IA + I(\lambda A) \subset I,$$

a contradiction.

We turn to the proof of Theorem 50. Let M be a faithful irreducible module for A. We know that M is isomorphic to A/I for a suitable regular maximal right ideal I. Of course, I does not contain A^2. It follows from the lemma that I is admissible under the centroid C, and hence so is A/I. The commuting division ring D of endomorphisms thus contains the centroid in a natural way (and also its quotient field).

Let $f = 0$ be the identity satisfied by A. We proceed to transform f into a multilinear homogeneous identity. If $f = f(x, \ldots)$ is not linear in x, write

$$g(u, v, \ldots) = f(u+v, \ldots) - f(u, \ldots) - f(v, \ldots).$$

Then g is satisfied by A, it is not the zero polynomial, and its degree in u or v is lower than the degree of f in x. By successive steps of this kind we reach a multilinear identity, whose joint degree in all its variables is still k; we shall again write it as f. Suppose f is not homogeneous. Then some variable, say x, is missing from at least one term. On setting $x = 0$ we get an identity of lower degree. Ultimately this will reach a homogeneous identity of degree $\leq k$. Changing notation again we assume the degree to be just k. Thus f consists of a term $x_1 x_2 \cdots x_k$ and some of its permutations. Consider the matrices $e_{11}, e_{12}, e_{22}, e_{23}, \ldots$. On substituting these into f, only one term survives, for in all other permutations the product of the e's is 0.

We now invoke the density theorem. A is a dense ring of linear transformations on a vector space V over D. If V is infinite-dimensional, it will be possible to find in A k elements which act exactly like $e_{11}, e_{12}, e_{22}, e_{23}, \ldots$ on a suitable finite-dimensional subspace of V. But this contradicts the identity f = 0.

Thus A is merely a total matrix ring over D. In particular it has a unit element, and its centroid C has become the ordinary center. We now regard D as an algebra over C, noting that it inherits the identity f = 0. Let K be a maximal subfield of D. In the algebra $K \otimes D$ the identity f = 0 survives, for a multilinear identity need only be checked on basis elements, and we can use a basis of D over C as a basis of $K \otimes D$ over K. By Theorem 47, $K \otimes D$ is a dense ring of linear transformations on a vector space over K. We now simply repeat the argument of the previous paragraph to deduce that this vector space is finite-dimensional. Hence A is finite-dimensional over C, say of dimension n^2. By suitable extension of the base field, we can suppose that A is merely an n by n total matrix algebra. Then k must be at least 2n, for if k = 2n-1, the use of the k elements $e_{11}, e_{12}, \ldots$ (just as before) would violate the identity.

We append a bibliography of the early work on polynomial identities, with short comments on each paper.

M. Hall, Projective planes, Trans. Amer. Math. Soc., vol. 54 (1943), 229-277. Theorem 6.2 states that if D is a division ring such that every $(xy - yx)^2$ is in the center, then D is one or four-dimensional over its center. This special case of Theorem 50 was its inspiration.

J. Levitzki, On a problem of A. Kurosch, Bul. Amer. Math. Soc. vol. 52 (1946), 1033-1035. Proof that a nil ring of bounded index is locally nilpotent.

I. Kaplansky, Rings with a polynomial identity, Bull. Amer. Math. Soc., vol. 45 (1948), 575-580. Foundations of the subject. Proof of Theorem 50. Also a proof that a nil ring satisfying a polynomial identity is locally nilpotent.

J. Levitzki, A theorem on polynomial identities, Proc. Amer. Math. Soc., vol. 1 (1950), 334-341. Let A be a ring with a polynomial identity of degree k. Let N be the union of all nilpotent ideals in A. Then every nilpotent element x in A satisfies $x^r \in N$, where $r = [k/2]$.

A. Amitsur and J. Levitzki, Minimal identities for algebras, Proc. Amer. Math. Soc., vol. 1 (1950), 441-463. Proof that the n by n total matrix ring over a commutative ring satisfies the "standard" identity $S_{2n} = 0$. Here is an interesting unpublished application. Let A and B be rings with unit elements. Suppose that A is commutative and that the n by n matrix rings A_n and B_n are isomorphic. Then B is commutative (whence A and B are isomorphic). For proof apply the standard identity to the $2n$ elements $\alpha e_{11}, \beta e_{11}, e_{12}, e_{22}, \ldots, e_{nn}$ in B_n to get $\alpha\beta = \beta\alpha$ for any α and β in B.

I. Kaplansky, Groups with representations of bounded degree, Can. J. of Math., vol. 1 (1949), 105-112. Application of polynomial identities to group representations.

A. Amitsur and J. Levitzki, Remarks on minimal identities for algebras, Proc. Amer. Math. Soc., vol. 2 (1951), 320-327. Determination of all identities of degree $2n$ for a simple algebra n^2-dimensional over its center.

I. Kaplansky, Topological representation of algebras II, Trans. Amer. Soc., vol. 68 (1950), 62-75. Proof that an algebraic algebra satisfying a polynomial identity is locally finite.

I. Kaplansky, The structure of certain operator algebras, Trans. Amer. Math. Soc., vol. 70 (1951), 219-255. Application of polynomial identities to C^*-algebras. Reduction of the local finiteness problem to the primitive and nil cases.

A. Amitsur, An embedding of PI-rings, Proc. Amer. Math. Soc., 3 (1952), 3-9. Let A be a ring with a polynomial identity and no nilpotent ideals. Then: (1) the degree of a minimal identity for it is even, (2) A can be embedded in a matrix ring over a commutative ring.

A. Amitsur, The identities of PI-rings, Proc. Amer. Math. Soc., 4 (1953), 27-34. Various further facts. In particular: any PI-ring satisfies an identity of the form $S_{2n}^m = 0$.

J. Levitzki, On the structure of algebraic algebras and related rings, Trans. Amer. Math. Soc., vol. 74 (1953), 384-409. Simplified purely algebraic proofs of local finiteness theorems. Many other results.

We conclude by mentioning an open question. An affirmative answer would have useful applications. Does the n by n matrix algebra $(n \geq 3)$ admit polynomials which are identically in the center without being identically 0?

(5) Extension of isomorphisms

THEOREM 51. Let E be a vector space over a division ring D with center Z. Let A be the ring of all linear transformations on E (note that the center of A is also Z). Let B and C be simple subalgebras of A, finite-dimensional over Z and containing the unit element of A. Then: any isomorphism between B and C can be extended to an inner automorphism of A.

COROLLARY. If A is itself finite-dimensional over Z, all its automorphisms are inner.

Proof. We operate in the algebra of all linear transformations on E as a vector space over Z. Among its subalgebras are D, A, B, C. By Theorem 45, $DB \cong D \otimes B$ is simple. By Theorem 48, $D \otimes B$ and $D \otimes C$ satisfy the descending chain condition. The given isomorphism ϕ between B and C extends in a natural way to an isomorphism, which we shall again call ϕ, between DB and DC. Now E is a right (DB)-module. There is another way of getting it to be a DB-module: by transferring to DC via ϕ, and then acting the way DC does on E. Explicitly, the second operation is defined by

$$a \cdot x = a\phi(x)$$

for a in E, x in DB.

Now each of these modules is a direct sum of irreducible modules, all isomorphic (for DB is simple with the descending chain condition). The two modules will be isomorphic as soon as we check that the number of irreducible components is the same both times. What we have is a decomposition of the vector space E into (DB)-submodules, which are in particular subspaces. The dimension of each subspace is finite (for an irreducible (DB)-module is isomorphic to a right ideal in DB, and all of DB is a finite-dimensional vector space over D). Now if a vector space is decomposed into finite-dimensional subspaces of a certain fixed dimension, the number of components is uniquely determined. Hence the two (DB)-modules are isomorphic. Call the isomorphism T. Then T is a one-one mapping of E onto itself satisfying

$$(ax)T = (aT)\phi(x)$$

for any x in DB. Apply this in particular with x in D, so that $\phi(x) = x$. The conclusion is that T is D-linear, that is, T lies in A. Next take x in B. We find $xT = T\phi(x)$. Hence on B, ϕ coincides with the inner automorphism by T. This concludes the proof of Theorem 51.

We shall outline several applications.

(a) The theorem of Frobenius. Let D be an algebraic division algebra over a real closed field R. We have already seen that D is R, the complexes over R, or four-dimensional over R. Let us complete the discussion of the last possibility. D contains the field $R(i)$, $i^2 = -1$. The automorphism $i \rightarrow -i$ can be extended to an inner automorphism of D, say by j. Then j^2 commutes with i and j, and so must be in the center R. Moreover j^2 cannot be positive, for then by ordinary factorization j would be in R. Hence j^2 is negative and we may normalize to $j^2 = -1$. It is easy to verify that the elements $1, i, j, ij$ are linearly independent and we reach the quaternions.

(b) Wedderburn's theorem. Let D be a finite division ring. Suppose its dimension over its center Z is n^2. Let K be a maximal subfield of D. Any other maximal subfield K_1, like K, has dimension n over Z. Hence K and K_1 admit an isomorphism leaving Z elementwise fixed. By Theorem 51 this extends to an inner automorphism of D. If we write K^* and D^* for the multiplicative groups of non-zero elements, we observe that the conjugates of K^* fill up D^*. But no finite group can be exhausted by a proper subgroup and its conjugates. Hence $K^* = D^*$, and D is commutative.

(c) Jacobson's generalization. Let D be an algebraic division algebra over a finite field. Let Z be its center. Let x be an element in D but not in Z. The field $Z(x)$ is normal over Z. Hence there exists an automorphism of $Z(x)$ over Z actually moving x into a polynomial $f(x)$. By Theorem 51 this extends to an inner automorphism, say by y. Consider now the subalgebra D_1 generated over Z by x and y. In view of the equation $y^{-1}xy = f(x)$, D_1 is finite-dimensional. By taking a basis of D_1, and dropping down to the finite field generated by the elements occurring in the multiplication table for that basis, we reach a finite division ring. There is a contradiction of Wedderburn's theorem unless D is commutative.

(d) Jacobson's theorem: if for every a in a ring A there exists an integer $n(a) > 1$ such that $a^{n(a)} = a$, then A is commutative. For one argues readily that the primitive images of A are division rings, and the latter are necessarily algebraic over a finite field.

Herstein made successive generalizations winding up with the following: suppose that for every a in A there exists a polynomial P_a with integral coefficients such that $a^2 P_a(a) - a$ is in the center of A; then A is commutative. Another theorem: if every element of A has some power in the center and A has no nil ideals, then A is commutative. For an authoritative account see Herstein's Carus Monograph Non-commutative Rings.

Part III. Homological Dimension

Introduction

These notes are based on a course given in the Autumn Quarter of 1958 and were written early in 1959.

The main objective of the course was to reach the Auslander-Buchsbaum-Serre characterization of regular local rings. At that time the result was still quite new, and its proof -- at least from a point reasonably near scratch -- was a sizeable undertaking.

Early in the course I formed a one-step projective resolution of a module, and remarked that if the kernel was projective in one resolution it was projective in all. I added that, although the statement was so simple and straightforward, it would be a while before we proved it. Steve Schanuel spoke up and told me and the class that it was quite easy, and thereupon sketched what has come to be known as "Schanuel's lemma". It took a couple of days and a half-dozen conversations before the proof was fully in hand.

Subsequently it became apparent that quite a few anticipations could be found in the literature. Most notably, Fitting (Math. Annalen 112 (1936), 572-582) had proved it with "projective" replaced by "free" and all modules finitely generated. However, Schanuel deserves full credit for stating it the right way and for realizing that it could lead to a theory of homological dimension (I will take a little credit for acting as a catalyst).

From this point on the course developed rapidly and took on the form recorded here. Many keen contributions were made by students, and I am especially grateful to H. Bass, S. Chase, and R. MacRae.

The course also contained two parts not reproduced here:
(1) An account of commutative Noetherian rings. An extended
version appears in the notes <u>Commutative Rings</u> issued by
Queen Mary College (QMC). (2) A similar theory of weak dimension. After the concept of flatness has been adequately
developed, one forms projective (or free, or flat) resolutions and
waits till the kernel is flat. An analogous sequence of theorems
can be worked up. I leave this as a long exercise to the interested
reader.

In the present reprinting there has been some editing and
some material has been incorporated from the QMC notes and
from my 1965 Varenna lectures.

1. Dimension of modules

R will always denote a ring with unit element, and all modules will be unitary. Normally, we shall deal with left modules.

We take the point of view that free modules are the simplest ones and we study other modules in terms of them. Of course, any module A is representable as the image of a free module F, say with kernel K:

$$(1) \qquad 0 \to K \to F \to A \to 0 \ .$$

The next simplest type of module is one for which K is free. But at once we face the question: is this independent of the particular choice of the resolution (1)? It turns out that the answer is "no", but becomes "yes" if we yield a little ground and replace "free" by "projective" (i.e., direct summand of a free module).

Once this is granted, it is natural to begin again, treating projective modules as the simplest type. The comparison of two projective resolutions

$$(2) \qquad 0 \to K \to P \to A \to 0 \ ,$$

$$(3) \qquad 0 \to K_1 \to P_1 \to A \to 0$$

is made in Theorem 1.

THEOREM 1. (Schanuel's Lemma). $\underline{\text{Let}}$ R $\underline{\text{be a ring,}}$ A $\underline{\text{an R-module, and let (2) and (3) be projective resolutions of}}$ A $\underline{\text{(i.e., the sequences are exact and}}$ P $\underline{\text{and}}$ P_1 $\underline{\text{are projective).}}$ $\underline{\text{Then}}$ $K \oplus P_1$ $\underline{\text{is isomorphic to}}$ $K_1 \oplus P.$

$\underline{\text{Proof.}}$ Let f, f_1 denote the maps from P, P_1 to A. Since P is projective there exists a map $g : P \to P_1$ with $f_1 g = f$. Let L denote the submodule of $P \oplus P_1$ consisting of the pairs (p, p_1) satisfying $f(p) = f_1(p_1)$. Map $P \oplus K_1$ into L by

$$(4) \qquad (p, k_1) \to (p, g(p) + k_1) \ .$$

It is straightforward that (4) is one-one and onto. Thus, if P is projective, $P \oplus K_1$ is isomorphic to L. If P_1 is also projective, L is isomorphic to $P_1 \oplus K$.

Theorem 1 inevitably suggests the introduction of an equivalence relation: modules A and B are equivalent if there exist projective modules P and Q such that $A \oplus P$ is isomorphic to $B \oplus Q$. We write $\mathcal{R}A$ for the equivalence class of K in (2), and the gist of Theorem 1 is that $\mathcal{R}A$ is well-defined. Furthermore we easily see that $\mathcal{R}A$ depends only on the equivalence class of A. We are now ready to define the projective dimension of A as the smallest n such that $\mathcal{R}^n A$ is the class of projective modules; if there is no such n, the projective dimension of A is ∞. We use the symbol $d(A)$, or $d_R(A)$ if it is advisable to call attention to the ring R.

Examples:

1. $d(A) = 0$ if and only if A is projective.

2. $d(A) = 1$ if and only if A is not projective but is expressible as B/C with B and C projective.

3. Let a, b be elements of R such that the left annihilator of a is Rb and the left annihilator of b is Ra. We have the exact sequence

$$0 \to Rb \to R \to Ra \to 0$$

where the map from Rb to R is inclusion, and that from R to Ra is right multiplication by a. We have a similar resolution with a and b interchanged. The result is to obtain a periodic long resolution that bounces back and forth between Ra and Rb. Thus $d(Ra) = d(Rb) = \infty$ unless Ra and Rb are both projective. When a and b are central, the condition for this is $(a, b) = R$. We give three illustrations:

(i) $R = Z_4$ (the integers mod 4), $a = b = 2$. Here $d(Ra) = \infty$.

(ii) Let u, v be central non-zero-divisors in a ring T, and suppose $(u, v) \neq T$. Let $R = T/(uv)$ and let a, b be the images of u and v in R. Again $d(Ra) = \infty$.

(iii) Let Y be any ring, $T = Y[x]$ with x a (commuting) indeterminate and set $u = x-1$, $v = 1 + x + \ldots + x^{n-1}$. Then $R = T/(uv)$ is the group ring over Y of the cyclic group of order n. The condition $(u, v) = T$ holds if and only if n is invertible in Y.

4. Exercise: if A is a direct sum of (any number of) modules B_i, then $d(A) = \sup d(B_i)$.

We proceed to a theorem giving a nearly complete relationship between the homological dimensions of three modules occurring in a short exact sequence. It is perhaps most useful to view this as an attempt to determine $d(A/B)$ from $d(A)$ and $d(B)$. This is successful except in the "ambiguous case" $d(A) = d(B)$ when we only get an inequality.

We call attention also to the condensed version suggested by P. M. Cohn: $d(A) \leq \max(d(B), d(C))$ with equality except possibly when $d(C) = d(B) + 1$.

THEOREM 2. Let B be a submodule of A, and write $C = A/B$; thus we have the exact sequence

(6) $$0 \to B \to A \to C \to 0.$$

(1) If two of the dimensions $d(A)$, $d(B)$, $d(C)$ are finite, so is the third.

(2) If $d(A) > d(B)$, then $d(C) = d(A)$.

(3) If $d(A) < d(B)$, then $d(C) = d(B) + 1$.

(4) If $d(A) = d(B)$, then $d(C) \leq d(A) + 1$.

<u>Proof</u>. Before beginning the proof we mention an alternate procedure, that of building simultaneous resolutions of all three modules (Cartan and Eilenberg, p.7, Prop. 2.5). Symbolically, we pass from (6) to

$$0 \to \mathcal{R}B \to \mathcal{R}A \to \mathcal{R}C \to 0.$$

Beginning the induction is a trifle more tedious in this version. Perhaps the best advice to the reader is to have both techniques at his disposal.

If A is projective, the theorem is immediate. If C is projective, then A is the direct sum of B and C and again the theorem is immediate. We may thus assume that neither A nor C is projective.

Write $A = P/D$ with P projective. Then B has the form E/D, where $D \subset E \subset P$, and $C \cong P/E$. Thus $d(E) = d(C) - 1$, $d(D) = d(A) - 1$, $d(E/D) = d(B)$. We have the exact sequence

$$0 \to D \to E \to B \to 0$$

or symbolically

(7) $$0 \to \mathcal{R}(A) \to \mathcal{R}(C) \to B \to 0 .$$

(It is interesting to note that two more applications of the procedure lead us to

$$0 \to \mathcal{R}^2 B \to \mathcal{R}^2 A \to \mathcal{R}^2 C \to 0).$$

By using (7) and induction on the sum of the two finite dimensions we get part one of the theorem at once. So we assume all three dimensions finite and make an induction on their sum. The inductive assumption on D, E, B gives the following information when translated back to A, B, C:

(a) If $d(C) > d(A)$, then $d(B) = d(C) - 1$,

(b) If $d(C) < d(A)$, then $d(B) = d(A)$,

(c) If $d(C) = d(A)$, then $d(B) \leq d(A)$.

These three statements are merely a logical rearrangement of the three statements in Theorem 2.

Exercise. All combinations permitted by Theorem 2 are actually possible.

2. Global dimension

The global dimension of R, written $D(R)$, is the sup of $d(A)$ taken over all R-modules. More exactly, this is the left global dimension and there is a similar right global dimension derived from right modules.

Brief arguments show that $D(R) = 0$ (left or right) if and only if R is semi-simple with descending chain condition. If R is an integral domain, $d(R) \leq 1$ if and only if R is a Dedekind ring.

If either of the two global dimensions is 0, so is the other. But otherwise there is no connection between the left and right global dimensions (A. V. Jatengaonkar, Notices Amer. Math. Soc. vol. 14, (1967), p. 660).

3. First theorem on change of rings

For many theorems or computations concerning homological dimension a comparison between two rings is useful. The following theorem, because it is so simple to prove and is decisive of its kind, deserves first mention.

THEOREM 3. <u>Let</u> R <u>be a ring with unit and</u> x <u>a central element of</u> R <u>which is a non-zero-divisor. Write</u> $R^* = R/(x)$. <u>Let</u> A <u>be a non-zero</u> R^*-<u>module with</u> $d_{R^*}(A) = n < \infty$. <u>Then</u> $d_R(A) = n + 1$.

The proper context in which to view results like Theorem 3 is the following: let R and S be rings and let there be given a ring homomorphism from R to S. Then any S-module A becomes in a natural way an R-module. By Theorem 2 and an easy induction one proves

$$(8) \qquad\qquad d_R(A) \leq d_S(A) + d_R(S).$$

(This is part of Exercise 5 on p. 360 of Cartan-Eilenberg). In Theorem 3 we have a case where the inequality (8) is improved to equality. By iterated use of Theorem 3 we can get further instances of equality (and this is a nice motivation for the concept of an R-sequence). In Corollary 2.12 of Auslander and Buchsbaum's <u>Codimension and multiplicity</u> (Ann. of Math. 68 (1958), 625-657) there is another case of equality. It would be interesting to know the precise circumstances in which equality holds.

<u>Proof of Theorem 3.</u> We proceed by induction on n.

$n = 0$. A is R^*-projective and hence a direct summand of a free R^*-module F. Now (x) is R-projective (even free on one generator) and not a direct summand of R. Hence $d_R(R^*) = 1$ and likewise $d_R(F) = 1$. It follows that $d_R(A) \leq 1$. We must exclude the possibility that A is R-projective. Now x acts faithfully on R, hence on any free R-module, hence on any non-zero submodule of a free R-module; therefore no non-zero projective R-module can be annihilated by x.

$n > 0$. Map a free R^*-module G onto A with kernel K. We have $d_{R^*}(K) = n-1$, whence $d_R(K) = n$ by induction. By Theorem 2 we can conclude $d_R(A) = n+1$ except when $n = 1$; here we only get $d_R(A) \leq 2$. (It is typical of such an inductive proof that the ambiguous case of Theorem 2 calls for a special argument at a low stage of the induction.) We shall conclude the proof by showing that when $d_R(A)$ and $d_{R^*}(A)$ are both at most one, A is R^*-projective.

Map a free R-module H onto A with kernel T. Since A is annihilated by x, $T \supset xH$. Since $d_R(A) \leq 1$, T is R-projective. We further have a homomorphism of H/xH onto A with kernel T/xH. Since H/xH is R^*-free and $d_{R^*}(A) \leq 1$, T/xH is R^*-projective. This implies that xH/xT is a direct summand of T/xT. Now T/xT is R^*-projective, as follows readily from the fact that T is R-projective. Hence xH/xT is R^*-projective. But $xH/xT \cong H/T \cong A$. Hence A is R^*-projective.

The next theorem is an immediate corollary of Theorem 3.

THEOREM 4. <u>Let</u> R, x, R^* <u>be as in Theorem 3. Suppose</u> $D(R^*) = n < \infty$. <u>Then</u> $D(R) \geq n+1$.

The following somewhat related theorem is due to D. E. Cohen.

THEOREM 5. <u>Let</u> T <u>be a subring of</u> R <u>and assume</u> T <u>is a direct summand of</u> R <u>as a</u> T-bimodule. <u>Then</u> $D(T) \leq D(R) + d_T(R)$.

<u>Proof</u>. Take any (left) T-module A, and set $B = \text{Hom}_T(R, A)$. Then B carries the structure of a left R-module. As a T-module it has A as a direct summand, for if $R = T \oplus U$ as a T-T-bimodule then $B = A \oplus \text{Hom}_T(U, A)$. It follows that $d_T(A) \leq d_T(B)$. By (8),

$d_T(B) \le d_R(B) + d_T(R)$. Hence $d_T(A) \le d_R(B) + d_T(R)$. Passing to global dimensions, we get the desired conclusion.

Useful examples for Theorem 5 are provided by taking R to be a polynomial algebra, power series algebra, group algebra, or semi-group algebra over R. Another example: let Y be a ring, G a group, H a subgroup of G, and take $T(R)$ to be the group ring over Y of $H(G)$. (The complement to T in R is provided by the set of linear combinations of elements not in H.)

4. Polynomial rings

Let S be any ring. By the polynomial ring $R = S[x]$ we mean the usual polynomials in x with coefficients in S; x commutes with the elements of S, but we allow S to be non-commutative. Note that x is in the center of R and is a non-zero-divisor.

THEOREM 6. Let $R = S[x]$ be a polynomial ring in x over S. Then $D(R) = 1 + D(S)$.

Proof. We have $D(R) \ge 1 + D(S)$, by Theorem 4 when $D(S) < \infty$ and (since $d_S R = 0$ here) by Theorem 5 when $D(S) = \infty$ (the argument below also looks after the case $D(S) = \infty$). We therefore assume $D(S) = n < \infty$ and have to prove $D(R) \le n + 1$.

Let A be any S-module. We describe a certain construction for a related R-module for which, at the moment, we write $A[x]$. (The reader who prefers is invited to substitute $R \otimes_S A$ and make analogous changes below.) The module $A[x]$ is the set of $\Sigma\, a_n x^n$, $a_n \in A$; the action of S and x (and thereby R) on $A[x]$ is self-explanatory.

We claim that $d_R(A[x]) = d_S(A)$. First: if A is S-free, evidently $A[x]$ is R-free. Direct summands offer no problem, so A S-projective implies $A[x]$ R-projective. Conversely, suppose $A[x]$ is R-projective. Then it is an R-direct-summand of an R-free module which is also S-free. (Observe that R is free as an S-module.) So $A[x]$ is S-projective. Since, as an S-module, $A[x]$ is merely a direct sum of a countable number of copies of A, it follows that A is S-projective.

Map a free S-module F onto A with kernel K. It is immediate that there is an induced R-homomorphism of $F[x]$ onto $A[x]$ with kernel $K[x]$; note also that $F[x]$ is R-free. We thus get parallel resolutions of A and $A[x]$ over S and $S[x]$ respectively, and it follows from the preceding paragraph that these resolutions terminate at the same moment. Moreover if one resolution never terminates the same is true of the other. We have sustained the claim that $d_S(A) = d_R(A[x])$.

We shall find it advisable later to change our notations for $A[x]$. We may write $A[x]$ instead as the set of all sequences $(a_0, a_1, a_2, \ldots)$ of elements of A, non-zero at only finitely many coordinates; S acts by pointwise multiplication and x as a push to the right:

$$x(a_0, a_1, a_2, \ldots) = (0, a_0, a_1, a_2, \ldots) .$$

Now let M be any R-module. We shall write N for the R-module which was described as $M[x]$ above; N is the set of all ultimately vanishing sequences $(m_0, m_1, m_2, \ldots)$, $m_i \in M$, with S acting pointwise and x acting as a push to the right. The map

$$(m_0, m_1, m_2, \ldots) \to \Sigma \, x^i m_i$$

defines an R-homomorphism of N onto M (note that the right side is meaningful since M is already an R-module). Let the kernel be K. We showed above that $d_R(N) = d_S(M)$; hence $d_R(N) \leq n$. We shall show further that K is isomorphic to N.

Application of Theorem 2 then yields that for $M = N/K$ we have $d_R(M) \leq n+1$, as desired.

The isomorphism of K and N is given by mapping N onto K as follows:

$$(m_o, m_1, m_2, \ldots) \rightarrow (xm_o, xm_1 - m_o, xm_2 - m_1, \ldots).$$

That this is indeed an R-homomorphism which is one-to-one and onto K is a straightforward verification that we leave to the reader.

Exercise. Adapt the above argument to the case $R = S[x, x^{-1}]$.

We record an immediate corollary of special interest.

THEOREM 7. (Hilbert's theorem on syzygies). If R is the ring of polynomials in n variables over a field, then $D(R) = n$.

5. Second theorem on change of rings

Theorem 3 does not suffice to get complete information on the connection between R and R^*, for the only R-modules covered are those annihilated by x. If in reverse we start with an R-module A, we need a way to pass to an appropriate R^*-module. The obvious choice is A/xA. With precautions about zero-divisors (which cannot be omitted) there is at any rate inequality:

THEOREM 8. Let R be a ring with unit, x a central element in R; write $R^* = R/(x)$. Let A be an R-module and suppose that x is a non-zero-divisor on both R and A. Then: $d_{R^*}(A/xA) \leq d_R(A)$.

Again there is a broader context in which to view Theorem 8. Given a ring homomorphism from R to S, and a (left) R-module A, we pass to $B = S \otimes_R A$ which is a (left) S-module. We are interested in the validity of $d_S(B) \leq d_R(A)$. If tensoring with S preserves exactness (i.e., if S is right R-flat), the inequality is immediate. This is not, however, the setup in Theorem 8, which is explained from the "higher" point of view as follows. The requisite preservation of exactness is assured if appropriate Tor's vanish. From Tor_2 on they do since $d_R(R^*) = 1$. The vanishing of $\text{Tor}_1^R(R^*, A)$ is precisely the hypothesis that x is a non-zero-divisor on A.

<u>Proof of Theorem 8.</u> If $d_R(A) = \infty$, there is nothing to prove. So we assume $d_R(A) = n < \infty$, and proceed by induction on n.

$n = 0$. Direct summands offer no problem so we may as well assume that A is R-free. Then A/xA is visibly R^*-free.

$n > 0$. Map a free R-module F onto A with kernel K. We have $d_R(K) = n-1$, whence $d_{R*}(K/xK) \leq n-1$ by induction. The map $F \to A$ followed by the natural homomorphism of A onto A/xA yields a map $F \to A/xA$ with kernel $K + xF$. We may instead regard this as a map $F/xF \to A/xA$ with kernel $(K + xF)/xF$. Now $(K + xF)/xF \cong K/(K \cap xF)$ by the standard isomorphism theorem. It follows readily from the hypothesis that x acts faithfully on A that $K \cap xF = xK$. Thus: we have a map of the R^*-free module F/xF onto A/xA with kernel K/xK. Since $d_{R*}(K/xK) \leq n-1$, we deduce $d_{R*}(A/xA) \leq n$, as desired.

6. Third theorem on change of rings

While Theorem 8 has a certain usefulness, it is to be expected that important results will concern the case of equality.

For this assumptions are needed concerning the Jacobson radical and finiteness.

THEOREM 9. Let R be a left Noetherian ring, x a central element lying in the Jacobson radical of R; write $R^* = R/(x)$. Let A be a finitely generated R-module. Assume that x is a non-zero-divisor on both R and A. Then $d_{R^*}(A/xA) = d_R(A)$.

Proof. Let $d_{R^*}(A/xA) = n$. We have to prove $d_R(A) = n$. If $n = \infty$, Theorem 8 applies, so we assume $n < \infty$. We are going to argue by induction on n. We do the inductive step first, leaving the discussion of $n = 0$ to the end.

Map a free finitely generated R-module F onto A with kernel K. This induces a map of F/xF onto A/xA with kernel K/xK (exactly as in the proof of Theorem 8). We have $d_{R^*}(K/xK) = n-1$ (since F/xF is R^*-free), whence $d_R(K) = n-1$ by induction. (Note that K satisfies the requisite conditions: it is finitely generated and x acts faithfully on it.) We conclude that $d_R(A) = n$. (If $n = 1$ there is a momentary possibility that $d_R(A) = 0$ but this of course implies $d_{R^*}(A/xA) = 0$.)

It remains for us to treat the case $n = 0$. That is, we must prove the following: under the hypotheses of Theorem 9, if A/xA is projective, then A is projective.

We first do this with "projective" replaced by "free" in both the hypothesis and the conclusion. Suppose then that A/xA is free over R^*, and let $v_1, \ldots, v_n$ be a basis. Pick elements u_i in A mapping on v_i. We claim that A is free, with $u_1, \ldots, u_n$ as a basis.

That the u's span A is typical deduction from Nakayama's lemma. In detail: let C be the submodule of A spanned by $u_1, \ldots, u_n$. We have $C + xA = A$, whence $x(A/C) = A/C$, and $A/C = 0$ by Nakayama.

Suppose $\Sigma\, c_i u_i = 0$ for $c_i \in R$. We show that the c's are 0. From $\Sigma\, c_i v_i = 0$ we get that each c_i is divisible by x. Since x acts faithfully on A, we may cancel x in the relation $\Sigma\, c_i u_i = 0$. The process may then be repeated. There will result a sequence of elements $c_i, c_i/x, c_i/x^2, \ldots$ which generates a properly ascending chain of left ideals in R, unless $c_i = 0$.

From the free case we pass to the projective case by a device due to Lance Small. Suppose then that A/xA is R^*-projective. Form a free resolution of A:

$$(9) \qquad 0 \to K \to F \to A \to 0 .$$

From (9) we pass to the corresponding resolution of A/xA:

$$(10) \qquad 0 \to K/xK \to F/xF \to A/xA \to 0 .$$

Let $B = A \oplus K$. Then B/xB is isomorphic to $A/xA \oplus K/xK$. Since A/xA is projective, the sequence (10) splits. Hence B/xB is isomorphic to F/xF, which is R^*-free. By the free case already treated, B is R-free, and A is R-projective as required.

There is an additional aspect of Theorem 9 which we shall briefly explore. Suppose, in the setup of Theorem 9, that $d_{R^*}(A/xA) = n < \infty$ and that $A \neq 0$ (whence, by Nakayama, $A/xA \neq 0$). Then, by Theorem 3, $d_R(A/xA) = n+1$. Putting this together with Theorem 9 we get

$$(11) \qquad d_R(A/xA) = 1 + d_R(A).$$

Now the interesting thing is that if R is commutative (11) can be improved in two respects: it is true also when $d_R(A) = \infty$, and we can delete the assumption that x is a non-zero-divisor in R. The proof is by the long exact sequence for Ext. Suppose $d_R(A/xA) = k < \infty$. Then $\mathrm{Ext}_R^{k+1}(A/xA, B) = 0$ for any R-module B.

From the exact sequence

$$0 \to A \xrightarrow{x} A \to A/xA \to 0$$

we get

$$\text{Ext}_R^k(A,B) \xrightarrow{x} \text{Ext}_R^k(A,B) \longrightarrow \text{Ext}_R^{k+1}(A/xA, B) = 0.$$

Take B finitely generated; then the Nakayama lemma yields $\text{Ext}_R^k(A,B) = 0$. This leads to $d_R(A) \leq k-1$, the key point we need.

Whether commutativity of R can be deleted here is unknown. A "little" non-commutativity can be allowed: if R is a T-algebra and a finitely generated module over the commutative Noetherian ring T, and x is in the Jacobson radical of T, then the above argument works (look at the relevant Ext as a T-module).

We wish to state a corollary of Theorem 9 applying to the global dimensions of R and R^*. Because of the restriction to finitely generated modules, there is a difficulty. In Section 14 we shall see that this difficulty is transitory, for the global dimension of a ring can be computed from its finitely generated modules, or even its cyclic modules. So: the distinction between D and $\overline{D}$ is temporary only.

DEFINITION. $\overline{D}(R)$ is the sup of $d(A)$, taken over all finitely generated modules.

THEOREM 10. _Let_ R _be a left Noetherian ring_, x _a central element in the Jacobson radical of_ R, _not a zero-divisor in_ R. _Let_ $R^* = R/(x)$. _Assume that_ $\overline{D}(R^*) = n < \infty$. _Then_ $\overline{D}(R) = n+1$.

Proof. That $\overline{D}(R)$ is at least $n+1$ is immediate from Theorem 3. Conversely, let A be any finitely generated R-module, say with $d_R(A) = k$. We must prove $k \leq n+1$. If $k = 0$, there is no problem. Otherwise we map a free finitely generated R-module F onto A with kernel K. We have $d_R(K) = k-1$.

Also, K is finitely generated and x acts faithfully on it. By Theorem 9, $d_R(K) = d_{R^*}(K/xK) \leq n$. Hence $k-1 \leq n$, $k \leq n+1$ as desired.

Exercise. Let S be a left Noetherian ring and R the formal power series ring in one (commuting) indeterminate over S. Prove: $\overline{D}(R) = 1 + \overline{D}(S)$.

7. Localization

We recall the fundamental definitions. R is a commutative ring with unit, S a multiplicatively closed subset containing 1. For any R-module A define A_S to be the set of pairs (a, s), $a \in A$, $s \in S$, with the identification $(a, s) = (a_1, s_1)$ if there exists $s_2 \in S$ with $s_2(s_1 a - sa_1) = 0$. With the usual rule of addition A_S is an abelian group. When this construction is performed on R, there is a natural multiplication making R_S a ring, and then A_S becomes an R_S-module in a natural way. Any R_S-module arises from an R-module this way (for instance from itself).

THEOREM 11. For any localization R_S of a commutative ring R, $D(R) \geq D(R_S)$, $\overline{D}(R) \geq \overline{D}(R_S)$. (The reader is reminded that the distinction between D and $\overline{D}$ is temporary.)

Proof. The proof is immediate from three simple remarks, whose proof we leave to the reader.

(1) If A is R-projective, then A_S is R_S-projective.

(2) If F is R-free and maps onto A with kernel K, then F_S is R_S-free and maps onto A_S with kernel K_S.

(3) A finitely generated R_S-module is of the form A_S with A a finitely generated R-module.

8. Preliminary lemmas

In sections 8-10 we derive the homological characterization of regular local rings. This section is devoted to some easy preliminary lemmas that are needed. R will denote a local ring (i. e., commutative, Noetherian, with unique maximal ideal M).

LEMMA 1. Let A be a finitely generated R-module, and B a direct summand of A such that $B \subset MA$. Then B = 0.

Proof. Say $A = B \oplus C$. We have $C + MA = A$, whence $M(A/C) = A/C$. By the Nakayama lemma, $A/C = 0$, whence B = 0.

LEMMA 2. Let A be a finitely generated R-module and $a_1, \ldots, a_n$ a minimal set of generators. Let F be a free R-module on n generators $u_1, \ldots, u_n$. Let F be mapped onto A by sending u_i into a_i. Then the kernel is contained in MF.

Proof. If $\Sigma c_i u_i$ is in the kernel and does not lie in MF then one of the c's, say c_1, must be a unit. From $\Sigma c_i a_i = 0$ we find that a_1 can be expressed in terms of the other a's, a contradiction.

LEMMA 3. Any finitely generated projective R-module A is free.

Proof. Map F on A via a minimal generation of A as in the preceding lemma. The kernel K is a direct summand of F. By Lemmas 1 and 2, K = 0, and A = F is free.

Remark. By auxiliary arguments it is possible to delete the hypothesis of finite generation in Lemma 3 (Ann. of Math. 68 (1958), 372-7).

LEMMA 4. Suppose that every element of M is a zero-divisor. Then for any finitely generated R-module A, d(A) is either 0 or ∞.

Proof. If not there exists a finitely generated A with $d(A) = 1$. Resolve A, $F \to A$, as in Lemma 2, so that the kernel K is projective (hence free) and satisfies $K \subset MF$. From the hypothesis that every element of M is a zero-divisor it is a known consequence that there exists a non-zero element z with $zM = 0$. But $zK = 0$ is impossible since K is free.

9. A regular local ring has finite global dimension

In connection with Theorem 13 we remind the reader that the distinction between D and $\overline{D}$ is a provisional one, to be removed in section 14.

THEOREM 12. _If_ R _is an_ n-_dimensional regular local ring,_ $\overline{D}(R) = n$.

Proof. Let M be the maximal ideal of R. Pick an element x in M but not in M^2 (this is possible except in the trivial case where R is a field). The element x is not a zero-divisor (indeed R is an integral domain). The ring $R^* = R/(x)$ is a regular local ring of Krull dimension $n-1$. By induction on the Krull dimension of R we may assume $\overline{D}(R^*) = n-1$. All the hypotheses of Theorem 10 are fulfilled and we conclude that $\overline{D}(R) = n$.

10. A local ring of finite global dimension is regular.

In proving the converse of Theorem 12 we are able simultaneously to extract the information that the one module M determines homological dimension.

THEOREM 13. <u>Let</u> R <u>be a local ring with maximal ideal</u> M. <u>Assume that</u> R <u>is not a field. Suppose</u> $d_R(M) = n < \infty$. <u>Then</u> R <u>is an</u> $(n+1)$-<u>dimensional regular local ring.</u>

<u>Proof.</u> We begin the proof by disposing of two rather trivial cases. Suppose $n = 0$, i.e., M is projective. By Lemma 3, M is free, which means that M is a principal ideal generated by a non-zero-divisor. From this it is easy to conclude that R is a one-dimensional regular local ring.

Suppose that every element in M is a zero-divisor. It then follows from Lemma 4 that $n = 0$, and this is covered by the preceding paragraph.

Let k be the Krull dimension of R. We shall argue by induction on k. If $k = 0$, every element of M is a zero-divisor (in fact even nilpotent); but this case we have already disposed of. We assume $k > 0$. We may of course further assume that M contains a non-zero-divisor. It is known that then there exists a non-zero-divisor x which is in M but not in M^2. Write $R^* = R/(x)$, and let $M^* = M/(x)$ be the maximal ideal of the local ring R^*. The Krull dimension of R^* is $k-1$.

It is a fact that M^* is isomorphic to a direct summand of M/xM. Supposing that this is known, let us see how the proof concludes. By Theorem 8, $d_{R^*}(M/xM) \leq d_R(M)$ and hence is finite. Therefore $d_{R^*}(M^*)$ is finite. By Theorem 2, $d_R(M^*) = d_R(M/(x)) = n$. (Note that (x) is projective, and that the ambiguous case of Theorem 2 does not arise since we have already taken care of the case $n = 0$.) By Theorem 3, $d_{R^*}(M^*) = n-1$; observe that it is vital to know somehow that $d_{R^*}(M^*)$ is finite. By induction on k, we have that R^* is an n-dimensional regular local ring. From this it follows that R is an $(n+1)$-dimensional regular local ring.

It remains to supply the proof that M^* is isomorphic to a direct summand of M/xM. Since $x \notin M^2$, we may pick a minimal base of M having the form $x, y_1, \ldots, y_r$. Let $S = xM + (y_1, \ldots, y_r)$. It is evident that $S + (x) = M$. Further, $S \cap (x) = xM$. For suppose $z \in S \cap (x)$. Then $z = ax = b_1 y_1 + \ldots + b_r y_r + cx$ $(a, b_i \in R, c \in M)$, and $ax - \Sigma b_i y_i \in M^2$. Since $\{x, y_i\}$ is a minimal base of M it follows that a lies in M, and thus $S \cap (x) = xM$. So M/xM is the direct sum of $(x)/xM$ and S/xM, and this implies that $M/(x)$ is isomorphic to a direct summand of M/xM, namely S/xM. This concludes the proof of Theorem 13.

Let R be a regular local ring, P a prime ideal in R. As is customary we write R_P instead of R_S, where S is the set-theoretic complement of P. Putting together Theorems 11, 12, and 13 we have:

THEOREM 14. <u>For any prime ideal</u> P <u>in a regular local ring</u> R, <u>the local ring</u> R_P <u>is again regular.</u>

11. Injective modules

In this section and the next three we develop the dual theory of injective dimension, and as a last step we apply it to obliterate the distinction between D and $\overline{D}$.

Let us begin by recalling that a module Q is injective if whenever modules $A \subset B$ and a homomorphism $f: A \to Q$ are given, f can be extended to B.

It is important to know that injectivity can be tested by just examining the case where B is the given ring.

LEMMA 5. <u>Let</u> Q <u>be a given</u> R-module. <u>Suppose that any homomorphism of a left ideal</u> I <u>into</u> Q <u>can be extended from</u> I <u>to</u> R. <u>Then</u> Q <u>is injective.</u>

See Cartan-Eilenberg, page 8, Theorem 3.2.

Next we have the dual of Theorem 1. The proof is dual and is left to the reader.

THEOREM 15. Suppose the sequences

$$0 \to A \to Q \to C \to 0$$

$$0 \to A \to Q' \to C' \to 0$$

are exact, and Q, Q' are injective. Then $Q \oplus C'$ is isomorphic to $Q' \oplus C$.

This suggests defining two modules to be <u>injectively equivalent</u> if they become isomorphic when suitable injective direct summands are added to each. The injective equivalence class of C in Theorem 15 is independent of the choice of the resolution and we write $\mathcal{J} A$ for it.

In order to build injective resolutions we need:

LEMMA 6. <u>Any module can be embedded in an injective module.</u>

A proof by Baer appears on page 9 of Cartan-Eilenberg; a proof by Eckmann and Schopf is sketched on page 31.

We now define the injective dimension of A to be the smallest n such that $\mathcal{J}^n A = 0$, ∞ if there is no such n. We introduce no symbol for injective dimension. For the global injective dimension of R (the sup of the injective dimensions of all R-modules) no symbol is needed, for it is equal to the global projective dimension. In order to prove this we must study Ext a little bit.

We conclude this section by noting the other characterization of injective modules. See Cartan-Eilenberg page 10, Lemma 3.4. (Remark: they use Lemma 6 but this can be avoided. Suppose Q has the universal direct summand property, let $A \subset B$, and suppose we are given $f: A \to Q$. Form $C = (B \oplus Q)/D$ where $D =$ all $(a, f(a))$. We get a one-to-one map of Q into C, and its splitting extends f to B. Compare the proof of Theorem 16.)

LEMMA 7. <u>A module is injective if and only if it has the property of being a direct summand of any module containing it.</u>

12. The group of homomorphisms

For any modules A and B we write $\text{Hom}(A, B)$ for the set of homomorphisms from A to B, made into an abelian group under the natural operation of addition. When R is commutative, $\text{Hom}(A, B)$ admits the structure of an R-module.

Given modules A, B, C and a map $A \to B$ there are natural induced maps $\text{Hom}(C, A) \to \text{Hom}(C, B)$ and $\text{Hom}(B, C) \to \text{Hom}(A, C)$.

LEMMA 8. <u>If $0 \to A \to B \to C$ is exact, then</u>

$$0 \to \text{Hom}(D, A) \to \text{Hom}(D, B) \to \text{Hom}(D, C)$$

<u>is exact. If</u> $A \to B \to C \to 0$ <u>is exact, then</u>

$$\text{Hom}(A, D) \leftarrow \text{Hom}(B, D) \leftarrow \text{Hom}(C, D) \leftarrow 0$$

<u>is exact.</u>

See Cartan-Eilenberg, page 26, Proposition 4.4.

13. The vanishing of Ext

We shall have no need to assign a meaning to Ext itself; we shall only speak of its vanishing.

DEFINITION. Let A, B be given modules. We say $\text{Ext}(A, B) = 0$ if the following is true: whenever a module C contains B with $C/B \cong A$ then B is a direct summand of C.

THEOREM 16. Let C, D be given modules. The following three statements are equivalent:

(1) $\text{Ext}(C, D) = 0$.

(2) For any exact sequence $0 \to A \to B \overset{g}{\to} C \to 0$ the sequence

(*) $\qquad 0 \leftarrow \text{Hom}(A, D) \leftarrow \text{Hom}(B, D) \leftarrow \text{Hom}(C, D) \leftarrow 0$

is exact.

(3) For a single exact $0 \to A \to B \to C \to 0$ with B projective, (*) is exact.

Proof. (1) $\to$ (2). Lemma 8 covers the exactness of (*) except at the term $\text{Hom}(A, D)$. So: we must prove that $\text{Hom}(B, D) \to \text{Hom}(A, D)$ is onto, i.e., we must show that any homomorphism h: $A \to D$ can be extended to a homomorphism $B \to D$. (We are thinking of A as being simply a submodule of B.) Let $E = D \oplus B$. Let T be the submodule of E consisting of all $(-h(a), a)$, $a \in A$. Write $F = E/T$. We have the sequence

(**) $\qquad\qquad 0 \to D \overset{p}{\to} F \overset{r}{\to} C \to 0$.

Here $p(d)$ is the class of $(d, 0)$ mod T, and $r(d, b) = g(b)$, this being independent of the choice of (d, b) within its class mod T. It is routine to check that (**) is exact. It follows from our hypothesis (1) that there exists a map $s: F \to D$ with $sp = $ identity. Define a map $B \to D$ in 3 steps

$$B \to E \to F \overset{s}{\to} D \ ,$$

the first being $b \to (0, b)$ and the second the natural homomorphism. One verifies that this map coincides with h when restricted to A.

(2) $\to$ (3). Trivial.

(3) $\to$ (1). Let E be a module with submodule D, $E/D \cong C$. We must prove that D is a direct summand of E. In the diagram

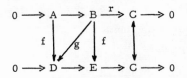

the map f arises since B is projective. When restricted to A, one easily sees that f maps into D. Our hypothesis (3) implies that $f: A \to D$ can be extended to $g: B \to D$. Now I define $t: E \to D$ as follows: pick $b \in B$ with $r(b) = s(e)$, and set $t(e) = e + g(b) - f(b)$. We have $t(e) \in D$ since $sf = r$, so that $st(e) = s(e) - sf(b) = 0$. If instead of b we take b_1 with $r(b_1) = s(e)$, then $g - b_1 \in A$ and $g - f$ vanishes on $b - b_1$; thus t is well-defined. Since g and f coincide on A, t is the identity on D. Hence D is a direct summand of E.

The gist of Theorem 16 is worth restating: to tell whether $Ext(A, B) = 0$, take a projective resolution of A:

$$0 \to K \to P \to A \to 0$$
$$\text{proj.}$$

and determine whether every homomorphism of K into B can be extended to P; the decision is independent of the choice of the resolution.

It is immediate from the definitions that $Ext(A, B)$ vanishes for all B if and only if A is projective and vanishes for all A if and only if B is injective. But with the aid of Lemma 5 this latter result can be usefully strengthened:

THEOREM 17. Let Q be an R-module such that $Ext(R/I, Q) = 0$ for every left ideal I in R. Then Q is injective.

The dual of Theorem 16 admits a dual proof that we leave to the reader.

THEOREM 18. <u>Let</u> C, D <u>be given modules. The following</u>
<u>three statements are equivalent.</u>

(1) Ext(C, D) = 0.

(2) <u>For any exact sequence</u> $0 \to D \to A \to B \to 0$ <u>the sequence</u>

(**) $0 \to \text{Hom}(C, D) \to \text{Hom}(C, A) \to \text{Hom}(C, B) \to 0$

<u>is exact.</u>

(3) <u>For a single exact sequence</u> $0 \to D \to A \to B \to 0$ <u>with</u>
A <u>injective,</u> (**) <u>is exact.</u>

14. Injective dimension

It is evident that the vanishing of Ext(A, B) depends only on
the projective equivalence class of A and the injective equivalence
class of B. Thus we may meaningfully speak of the vanishing of
Ext($\mathcal{R}$ A, B) or Ext(A, $\mathcal{I}$ B). It turns out that these two statements
are equivalent.

THEOREM 19. <u>For any modules</u> A <u>and</u> B, Ext($\mathcal{R}$A, B) = 0
<u>if and only if</u> Ext(A, $\mathcal{I}$B) = 0.

<u>Proof.</u> We shall suppose Ext($\mathcal{R}$ A, B) = 0 and prove
Ext(A, $\mathcal{I}$ B) = 0; the other half of the proof is dual.

Take a projective resolution of A and an injective resolution of B:

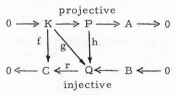

We must prove Ext(A, C) = 0. By Theorem 16 this means that we
must take any f: K → C and prove that f can be extended to P.
By Theorem 18, the hypothesis Ext(K, B) = 0 implies that there

exists $g: K \to Q$ with $rg = f$. Since Q is injective, g can be extended to $h: P \to Q$. Then rh is the desired map of P into C.

THEOREM 20. <u>The projective and injective global dimensions of any ring are equal.</u>

<u>Proof.</u> Let n be the projective global dimension. We shall show that the injective global dimension is at most n; the other half is dual. The case $n = \infty$ being trivial, we assume n finite. We must show $\mathcal{I}^n B = 0$ for any B, i.e., $\mathrm{Ext}(A, \mathcal{I}^n B) = 0$ for any A, B, i.e., (by n successive applications of Theorem 19) $\mathrm{Ext}(\mathcal{R}^n A, B) = 0$, which is true by hypothesis.

At last we supply the proof that D and $\overline{D}$ coincide.

THEOREM 21. <u>The projective global dimension of any ring R is the sup of $d(A)$ taken over all cyclic modules A.</u>

<u>Proof.</u> Given $\mathcal{R}^n R/I = 0$ for any I we must prove $\mathcal{R}^n C = 0$ for any module C. That is, we must show $\mathrm{Ext}(\mathcal{R}^n C, B) = 0$ for any C, B, i.e., $\mathrm{Ext}(C, \mathcal{I}^n B) = 0$, i.e., $\mathcal{I}^n B$ injective, i.e., (Theorem 18) $\mathrm{Ext}(R/I, \mathcal{I}^n B) = 0$, i.e., $\mathrm{Ext}(\mathcal{R}^n R/I, B) = 0$, which is true by hypothesis.

NOTES

Page 9.   To the three classical ruler and compass problems a fourth should be added: the construction by ruler and compass of a regular polygon of n sides. This is equivalent to asking whether the number $u = \cos(2\pi/n)$ is a constructible real number. Write $\varepsilon = e^{2\pi i/n}$, a primitive n-th root of unity. Then $\varepsilon + \varepsilon^{-1} = 2u$. It follows readily that the degree of ε over the field Q of rational numbers is twice the degree of u over Q. Let us write $g(n)$ for the degree of ε over Q. Then we see that a <u>necessary</u> condition for $\cos(2\pi/n)$ to be constructible is that $g(n)$ is a power of 2. In fact, this condition is also sufficient, but a proof at this point is not easily given, so we delay it till later (see the section of these notes referring to page 33).

What is $g(n)$? The answer is known; $g(n)$ is equal to the Euler function $\phi(n)$, the number of residue classes mod n which are prime to n. This fact (which can be restated as the irreducibility of the so-called cyclotomic polynomial) is somewhat tricky to prove and we shall not discuss the general proof, which can be found in several of the available treatises on modern algebra. However, the case where n is a prime (say p) is comparatively easy. The problem is to prove that

$$\frac{x^p - 1}{x-1} = x^{p-1} + x^{p-2} + \ldots + x + 1$$

is irreducible over the field of rational numbers. We set $x = y+1$, and find that the polynomial becomes

$$\frac{(y+1)^p - 1}{y} = y^{p-1} + py^{p-2} + \binom{p}{2}y^{p-3} + \ldots + p .$$

In this new polynomial the highest coefficient is 1, every subsequent coefficient is divisible by p, and the constant term is not divisible by p^2. By Eisenstein's criterion, the polynomial is irreducible. (This discussion is borrowed from Birkhoff and Mac Lane.)

So now we have to study a problem in number theory: given an odd prime p, when is it true that p-1 is a power of 2, say 2^n ? We summarize the facts. The number n in turn has to be a power of 2, say 2^t. The numbers $F_t = 1 + 2^{2^t}$ are called Fermat numbers, honoring Fermat, who thought they were all prime. The first five are 3, 5, 17, 257, and 65537 (corresponding to $0 \leq t \leq 4$), and are indeed prime. But for $t \geq 5$, F_t has turned out to be composite in every case which has been decided. For a survey of the status of the Fermat numbers, see Wrathall [19] (these numbers refer to the bibliography at the end of the notes). The factorization of F_7 was accomplished very recently [10], although F_7 had been proved to be composite by Morehead and Western in 1905.

It is a simple matter to extend the investigation to composite n. It turns out that $\phi(n)$ is a power of 2 if and only if n has the following form: a power of 2 multiplied by a product of distinct Fermat primes.

Page 13. There is a very nice theorem, due to Barbilian [1], which in a way rounds out the circle of ideas presented up to this point. Krull [9] simplified the proof and developed the theory further.

Here is the theorem. <u>Let</u> K $\subset$ L <u>be fields and let</u> G <u>be the Galois group of</u> L/K. <u>Assume that the Galois correspondence between subgroups of</u> G <u>and intermediate fields is absolutely perfect.</u> <u>Then</u> L <u>must be finite-dimensional over</u> K (<u>and of course normal</u>).

Suppose that instead of requiring a perfect correspondence between all intermediate fields and all subgroups, we yield some ground by assuming that the Galois group is topologized and that only closed subgroups are eligible. Then in the case of normal algebraic extensions we do again get a perfect correspondence, as is mentioned on page 77. With transcendental extensions allowed, new problems arise. They are studied in [17], where earlier references are also given.

Page 33. Right after Lemma 3 is a convenient place to break in and complete the discussion of ruler and compass construction of regular polygons. In the notes on page 9 we proved the "only if" portion of the following theorem. Now it is quite routine to supply the "if" part.

THEOREM. Let p be an odd prime. Then a regular polygon of p sides is constructible by ruler and compass if and only if p is a Fermat prime.

Granted the irreducibility of the cyclotomic polynomial, one gets the complete result.

THEOREM. A regular polygon of n sides is constructible by ruler and compass if and only if n has the following form: a power of 2 multiplied by a product of distinct Fermat primes.

The following additional information is worth recording. Let Q be the field of rational numbers, let u be a real algebraic number, and let K be a normal closure of $Q(u)$ over Q. Then: u is constructible by ruler and compass if and only if $[K:Q]$ is a power of 2. To prove this, one needs routine arguments combined with the following theorem from group theory: a finite group with order a power of 2 has a non-trivial center.

Pages 40-42. Theorems 31 and 33 are special cases of a broader theory. In the text, this broader theory was not developed since applications were envisaged only for Theorems 31 and 33.

Here is a sketch of the more general results. Let L be normal and finite-dimensional over K, with Galois group G. Then G acts on two abelian groups: L^+, the additive group of L, and L^*, the multiplicative group of non-zero elements in L. Now any time a group G acts on an abelian group A, certain cohomology groups $H^n(G,A)$ are definable. Theorems 31 and 33 say that $H^1(G, L^+)$ and $H^1(G, L^*)$ vanish if G is cyclic. Actually, $H^1(G, L^*) = 0$ for any G and $H^n(G, L^+) = 0$ for any G and any $n \geq 1$. The cohomology group $H^2(G, L^*)$ does not in general vanish; it connects with the Brauer group of K. Two references for further reading are Serre [14], [15].

Page 71. The idea in Theorem 63 has been developed further by Isaacs [5].

Page 102. In collaboration with Adjan, Novikov has now published full details of his work on the Burnside problem. In three papers [11] they prove that the non-trivial Burnside groups with odd exponent ≥ 4381 are infinite. In a further paper [12] they show that the groups in question are not even finitely presented.

Page 123. The problem about the coefficient of the unit element in an idempotent has been ingeniously resolved in the affirmative by A. Zalessky (letter to the author). He also proves the analogous result in characteristic p; in fact, he does it first for characteristic p and then succeeds in making a reduction mod p.

Let me mention at this point the publication of Passman's comprehensive monograph [13] on infinite-dimensional group algebras.

Page 161. The question on central polynomials has been answered affirmatively by Formanek [4].

Page 166. The Queen Mary College notes are now out of print. In nearly all respects, they have been superseded by [8].

Page 171. The paper [6] contains full details on the result announced by Jategaonkar.

Page 172. The first change of rings theorem (Theorem 3) has received some attention in the literature. Pertinent references are [2], [3], [7], and [16]. Injective analogues of the three change of rings theorems are established in Section 4-4 of [8].

Page 178. The proof that begins in the second last paragraph (it is due to Bass) can be replaced by the following argument, which proves a little more.

PROPOSITION. Let x be a central element in the Jacobson radical of R. Suppose that x is a non-zero-divisor on a finitely presented R-module A, and that A/xA is R^*-free, where $R^* = R/(x)$. Then A is R-free.

Proof. Again take $v_1, \ldots, v_n$ to be a basis of A/xA, lift v_i to $u_i \in A$, and observe à la Nakayama that the u's span A. Resolve A over R:

$$0 \to K \to F \to A \to 0.$$

Here F is free on $z_1, \ldots, z_n$, and z_i maps to u_i. If $\Sigma t_i z_i \in K$, then $\Sigma t_i^* v_i = 0$, where t_i^* is the image of t_i in R^*. It follows that $t_i^* = 0$, whence $t_i = x s_i$. We deduce $x \Sigma s_i u_i = 0$. Since x is a non-zero-divisor on A, we have $\Sigma \, s_i u_i = 0$ and $\Sigma \, s_i z_i \in K$. In short, $K = xK$. Now K is finitely generated since A has been assumed to be finitely presented. By a second application of Nakayama, $K = 0$, as desired.

For another way of proving Theorem 9, see the paper [18] by Strooker.

Pages 183-5. For the possible convenience of some readers I will indicate (by theorem number) where some needed facts can be found in [8]. (Of course they are in lots of other books as well.)

On page 183, R is given as a regular local ring. Then R is a domain (Theorem 164). With x in $M - M^2$, $R/(x)$ is regular (Theorem 161) and $(n-1)$-dimensional (Theorem 159).

On page 184, granted that the maximal ideal M does not consist of zero-divisors, we need a non-zero-divisor x in $M - M^2$. This is provided by Theorem 83, in conjunction with Theorem 80. Finally, granted that $R/(x)$ is a n-dimensional regular local ring, we use Theorems 162 and 84 to deduce that R is an $(n+1)$-dimensional regular local ring.

BIBLIOGRAPHY FOR THE NOTES

1. D. Barbilian, Solutia exhaustiva a problemai lui Steinitz, Acad. Repub. Pop., Romǎne. Stud. Cerc. Mat. 2(1951), 195-259. Romanian with French and Russian summaries. Reviewed in Mathematical Reviews 16, 669.

2. J. Cohen, A note on homological dimension, J. of Alg. 11(1969), 483-7.

3. K. Fields, On the global dimension of skew polynomial rings, J. of Alg. 13(1969), 1-4.

4. E. Formanek, Central polynomials for matrix rings, to appear in J. of Alg.

5. I. M. Isaacs, Degrees of sums in a spearable field extension, Proc. Amer. Math. Soc., 25(1970), 638-641.

6. A. Jategaonkar, A counter-example in ring theory and homological algebra, J. of Alg. 12(1969), 418-440.

7. C. Jensen, Some remarks on a change of rings theorem, Math. Zeit. 106(1968), 395-401.

8. I. Kaplansky, Commutative Rings, Allyn and Bacon, 1970.

9. W. Krull, Über eine Verallgemeinerung des Normal-körperbegriffs, J. Reine Angew. Math. 191(1953), 54-63.

10. M. Morrison and J. Brillhart, The factorization of F_7, Bull. Amer. Math. Soc. 77(1971), 264.

11. P.S. Novikov and S.I. Adjan, Infinite periodic groups, I, II, III, Izv. Akad. Nauk SSSR Ser. Mat. 32(1968), 212-244, 251-524, and 709-731; translations Math. USSR Izv. 2(1968), 209-236, 241-279, and 665-685.

12. ———— , Defining relations and the word problem for free periodic groups of odd order, same Izv. 32(1968), 971-979; translation 2(1968), 935-942.

13. D. Passman, Infinite Group Rings, Dekker, 1971.

14. J. -P. Serre, Cohomologie Galoisienne, Lecture Notes in Mathematics no. 5, Springer, 1964.

15. ———— , Corps Locaux, 2-nd ed. , Hermann, Paris, 1968.

16. L. Small, A change of rings theorem, Proc. Amer. Math. Soc. 19(1968), 662-6.

17. T. Soundararajan and K. Venkatachaliengar, On Krull Galois theory for non-algebraic extension fields, Bull. Austral. Math. Soc. 4(1971), 367-387.

18. J. Strooker, Lifting projectives, Nagoya Math. J. 27(1966), 747-751.

19. C. Wrathall, New factors of Fermat numbers, Math. Comp. 18(1964), 324-325.

INDEX